T0142104

Terrorism, Security, and Computation

Series Editor

V. S. Subrahmanian, College Park, USA

For further volumes:
http://www.springer.com/series/11955

V. S. Subrahmanian · Aaron Mannes
Animesh Roul · R. K. Raghavan

Indian Mujahideen

Computational Analysis and Public Policy

Foreword by The Hon. Louis J. Freeh

Springer

V. S. Subrahmanian
Department of Computer Science
University of Maryland
College Park, MD
USA

Aaron Mannes
Institute for Advanced Computer Studies
University of Maryland
College Park, MD
USA

Animesh Roul
Society for the Study of Peace and Conflict
Dwarka, Delhi
India

R. K. Raghavan
Central Bureau of Investigation
Alwarpet, Chennai
Tamil Nadu
India

ISSN 2197-8778 ISSN 2197-8786 (electronic)
ISBN 978-3-319-37844-2 ISBN 978-3-319-02818-7 (eBook)
DOI 10.1007/978-3-319-02818-7
Springer Cham Heidelberg New York Dordrecht London

Softcover reprint of the hardcover 1st edition 2013

Printed on acid-free paper

Springer is part of Springer Science+Business Media (www.springer.com)

Foreword

"Those who cannot remember the past are condemned to repeat it."
George Santayana, The Life of Reason.

Santayana's advice though spoken in the political sphere is grounded in the age-old military doctrine of knowing your enemy in order to defeat him. Such sage thinking has been repeated from Sun Tzu circa 500 BC to present professors of military tactics today.

Knowing your enemy is an inherent core of this timely and important book on the Indian Mujahideen, a terrorist group, by a foursome of respected scholars (R. K. Raghavan, Animesh Roul, Aaron Mannes, and V. S. Subrahmanian). The writers have successfully revealed the inside workings of this dangerous terrorist group, which is presently a major threat to the Country of India. By examining the origins, known supporters, and current tactics of the IM, this well-organized book succeeds in letting India gain a deep knowledge of the enemy targeting the national security of India and its economic prosperity.

We have learned in the fight against terrorism that the organizations who sponsor and finance these crimes against humanity have 'different faces.'

For example, when the self-styled 'Saudi Hezbollah' used a massive truck bomb to destroy Khobar Towers in June 1996, where USAF personnel enforcing the 'no fly' zone over southern Iraq were billeted, murdering 18 Airmen and injuring hundreds, the real terrorists behind the deadly attack was the Iranian IRGC. As with the IM, the real 'principals' behind these highly organized and intricately planned attacks have to be identified and brought to justice, whatever the political and foreign policy considerations.

The book is extremely well researched with copious details from the beginnings of the IM to its present day operations and policies. It is most thoughtful, especially in the chapter dealing with the IM's modus operandi. The portions of the book, most impressive and insightful, are those which focus on approaches to analyze the IM's past attacks as possibly predictive of future IM terrorist operations. The true strength of the book, however, is its recommendation for the establishment of a robust National Counter-Terrorism Center in India. India needs to step forward to an effective long-term solution not just to fight the clear and present danger of the IM, but to face the realities of the global terrorism challenge that all nations face.

In this important regard, I highly recommend this significant and well-written book.

Director, Federal Bureau of Investigation (1993–2001), US Louis J. Freeh

Advance Praise

"This book presents a highly innovative computational approach to analyzing the strategic behavior of terrorist groups and formulating counter-terrorism policies. It would be very useful for international security analysts and policymakers." *Uzi Arad, National Security Advisor to the Prime Minister of Israel and Head, Israel National Security Council (2009–2011).*

"An important book on a complex security problem. Issues have been analysed in depth based on quality research. Insightful and well-balanced in describing the way forward". *Naresh Chandra, Indian Ambassador to the USA (1996–2001) and Cabinet Secretary (1990–1992).*

"An objective and clinical account of the origins, aims, extra-territorial links and modus-operandi, of a growingly dangerous terrorist organization that challenges the federal, democratic, secular and pluralistic ethos of India's polity. The Authors have meticulously researched and analysed the multi-faceted challenges that the "*Indian Mujahideen*" poses and realistically dwelt on the ways in which these challenges could be faced and overcome." *G. Parthasarathy, High Commissioner of India to Australia (1995–1998) and Pakistan (1998–2000).*

"The book is a commendable effort to educate us about the IM. Its well researched and contains information that should prove useful for professionals dealing with terrorism such as details of the organization, computational predictions and the concept of Temporal Probabilistic Behavior Rules. The authors need to be complimented for attempting a mathematical definition of a policy against the IM, for a sound analysis of the IM and for offering some very sound operational suggestions. It is particularly creditable that the book, even though it deals with such a serious subject, is eminently readable by lay persons and therefore should educate them on a subject that is extremely relevant. Is a valuable addition to material that helps in our fight against terrorism. The authors deserve to be applauded for this authoritative book". Nikhil Kumar, Governor of Kerala (March 2013–to date), ex-Governor of Nagaland (Oct 2009–March 2013), ex-Director General of India's National Security Guards (June 1999–July 2001), ex-Commissioner Delhi Police (Jan 1995–April 1997).

Preface

Three recent advances have revolutionized the study of the behaviors of both individuals and groups. First, the wide and near-instantaneous dissemination of information on the Internet has led to a plethora of data about terrorist groups as well as the actions of other actors that might influence these groups. Second, text analysis techniques of the kind used by web search engines are significantly enhancing our ability to search this plethora of data. Third, huge advances in data mining allow us to track hundreds of variables about terrorist groups and use them to create models that generate factors predicting terrorist attacks.

The Laboratory for Computational Cultural Dynamics (LCCD) at the University of Maryland was created in 2004 to study the behavior of terrorist groups using computational methods, to develop systematic, data-driven methods that predict the future behaviors of these groups, and to identify policies that reduce the level of terrorist acts carried out by these groups. LCCD consists of an interdisciplinary research team including, over the years, computer scientists, political scientists, psychologists, public policy experts, criminologists, and sociologists. We emphasize the interdisciplinary nature of behavioral studies—and to reinforce this approach, this book has four Authors. One is a Computer Scientist (Subrahmanian) with a 10-year background in counter-terrorism research, another (Mannes) is a Public Policy Expert with over 10 years of experience in counter-terrorism, the third (Roul) is a Policy Analyst with nearly 15 years of experience in terrorism and counter-insurgency research, while the last (Raghavan) is a Political Scientist and Criminologist with nearly 50 years of experience as a law enforcement expert.

LCCD started with small-scale studies of Hamas and Hezbollah and then undertook the first comprehensive study of individual terrorist groups using computational means, co-authored by the first two Authors of this book in 2012 along with others. The result was a book, *Computational Analysis of Terrorist Groups: Lashkar-e-Taiba*, by Subrahmanian, Mannes, Sliva, Shakarian, and Dickerson, published by Springer in 2012 (Subrahmanian et al. 2012). After that book went to press, LCCD put out several forecasts regarding LeT's behavior and those of proxy groups associated with it; most of the forecasts have come true to date. These predictions are summarized in the table below (See Table 1).

Table 1 LCCD predictions and outcomes, LeT

	Date predicted	Predicted time frame	Prediction	Outcome
1	May 2011, counter-terrorism meeting in the Washington DC area	July–December 2011	Armed clashes between LeT/LeT proxies and Indian security forces	Several such incidents occurred
2			Civilians will be abducted by LeT/LeT proxies	Several such incidents occurred
3			LeT/LeT proxies will let some abducted civilians go	Several civilians abducted by LeT "escaped"; Unclear if LeT let them go intentionally
4			Armed clashes between LeT/LeT proxies and other armed groups	LeT targeted leaders of the JKLF[a]
5			Fedayeen attacks on production sites will be carried out by LeT and proxies	None reported though several attacks were thwarted
6	December 2012, Dutch police conference; January 2013, Aspen Institute Delhi talk	January–March 2013	Predicted no major LeT/LeT proxy attacks on either professional security forces or on public, symbolic, transportation sites	No such attacks occurred
7			Predicted that LeT/LeT proxies will carry out small attacks on civilians	Several such attacks occurred
8	May 2013, U.S. Consulate talk in Mumbai	June–September 2013	Predicted that there will be LeT attacks on either professional security forces or public, symbolic, or transportation targets	The Bodh Gaya attacks in June 2013 are believed to have been carried out either by LeT or IM[b]
9			Predicted that there will be small LeT attacks on civilians	Still to be determined

[a] Jammu Kashmir Liberation Front
[b] Indian Mujahideen

According to Table 1, of the eight predictions that can be evaluated at this time, only one turned out wrong (Prediction 5); all the others were largely correct. The correctness of Prediction 9 will only be clear by the end of September 2013.

Because of the close relationship between Lashkar-e-Taiba (LeT) and the Indian Mujahideen (IM), LCCD collected data on IM during that study. IM, which largely carries out terrorist attacks within the borders of India, has the potential to destabilize India. It deserves to be studied in its own right. This book presents the first comprehensive analysis of IM based on a robust computational and mathematical framework. Our work builds on excellent prior studies, notably that of Shishir Gupta (2011).

We owe a debt of gratitude to many people who shaped our thinking about IM. While working for LCCD, Jana Shakarian collected a first set of data about IM, which we then adapted, extended, and improved. We have benefited from conversations about IM with many people in India. On terrorism in South Asia, we have benefited from discussions with Marvin Weinbaum and Stephen Tankel. On terrorism in general, we have learned much from discussions with Daveed Gartenstein-Ross, Roy Lindelauf, and Jonathan Wilkenfeld. Conversations between one of the Authors (Raghavan) and several current and past senior law enforcement and intelligence officers in India were helpful in checking and crosschecking some facts. This exercise, we believe, has greatly enhanced the credibility of this study.

On the technology side, the Stochastic Temporal Analysis of Terrorist Events (STATE) system used in this book was primarily developed by two University of Maryland undergraduates, Moshe Katz and Yehuda Katz, who built it on top of prior work on a temporal probabilistic (TP) rule-mining engine. This algorithm was originally developed by Jason Ernst (now at UCLA) and the first Author, and later improved significantly by LCCD member John Dickerson (now at Carnegie Mellon). The TP-rules themselves were originally developed by Alex Dekhtyar and V. S. Subrahmanian. The use of mixed integer programming methods to generate policies was based on work by the first Author with his then-PhD student Raymond Ng (now at the University of British Columbia), Colin Bell (now at Microsoft), and Anil Nerode (at Cornell).

The work of LCCD on developing the technology used in this analysis has been generously funded by many organizations, primarily the Army Research Office (ARO) under grants W911NF1110344, W911NF0910206, and W911NF11C0215 and NSF grant SES0826886. In particular, Dr. Purush Iyer at ARO has been a tireless friend, supporter, and constructive critic of our work.

On the administrative side, Barbara Lewis helped with formatting and checking the manuscript and Noseong Park helped with some figures. Althea Nagai and Michael Garber proofed and critiqued the manuscript in its near-final stage. John Dickerson ran some of the computational tools required to extract the rules.

College Park, MD, August 25, 2013 V. S. Subrahmanian
Aaron Mannes
New Delhi, India Animesh Roul
Chennai, India R. K. Raghavan

Reference

Subrahmanian VS, Mannes A, Shakarian J, Sliva A, Dickerson J (2012) Computational analysis of terrorist groups: Lashkar-e-Taiba. Springer, New York

Contents

Acronyms

ACE	Automated Coding Engine
ANFO	Ammonium Nitrate and Fuel Oil
APT	Annotated Probabilistic Temporal
ARCF	Asif Reza Commando Force
BJP	Bharatiya Janata Party
CBI	Central Bureau of Investigation
CCTNS	Crime and Criminal Tracking and Network Systems
CID	Central Investigation Division
CPMF	Central Para-Military Forces
FARC	Fuerzas Armadas Revolucionarias de Colombia
FDLR	Forces Democratiques de Liberation du Rwanda
GCHQ	Government Communications Headquarters
GMRF	Gujarat Muslim Revenge Force
HM	Hizbul Mujahideen
HuJI	Harkat-ul-Jihad-i-Islami
HuM	Harkat-ul-Mujahideen
IB	Intelligence Bureau
IED	Improvised Explosive Device
IJT	Islami Jami'at-i-Tulaba
IM	Indian Mujahideen
ISI	Inter-Services Intelligence
JeI	Jamaat-e-Islami
JeM	Jaish-e-Mohammed
JIH	Jamaat-e-Islami Hind
JIJK	Jamaat-e-Islami Jammu and Kashmir
JTFI	Joint Task Force on Intelligence
LeT	Lashkar-e-Taiba
MAC	Multi-Agency Centre
MHA	Ministry of Home Affairs
NCTC	National Counter-Terrorism Center
NGO	Non-Governmental Organization
NIA	National Investigation Agency
NSAG	Non-State Armed Group
OASYS	Opinion Analysis System

PAGE	Policy Analytics Generation Engine
PCA	Policy Computation Algorithm
PLO	Palestine Liberation Organization
R&AW	Research and Analysis Wing
SIMI	Students Islamic Movement of India
SIO	Students Islamic Organization of India
SMS	Short Message Service
SOMA	Stochastic Opponent Modeling Agent
STATE	Stochastic Temporal Analysis of Terrorist Events
TIM	Tanzim Islahul Muslimeen
TOSCA	Trie-Oriented Stage Change Attempts
TP	Temporal Probabilistic
UK	United Kingdom
ULFA	United Liberation Front of Assam
USIS	United States Information Service

Chapter 1
Introduction

Abstract The Indian Mujahideen (IM) is a terrorist group that seeks to destabilize India and retaliate for the perceived mistreatment of Muslims in India. Though its origins are unclear and the subject of some debate, there is no question that the group has been responsible for about 750 deaths since October 2005. We analyzed over 770 variables relating to IM on a monthly basis for a 9 year period, from 2002 to 2010, in order to identify the conditions under which IM carries out different kinds of attacks. Using sophisticated "big data" analytics methods, we were able to derive over 25,000 rules relating to IM's behavior, of which roughly 29 are presented in this book. In addition, using sophisticated "policy generation" techniques, we were also able to come up with policy recommendations that have the potential of reducing attacks by IM.

On a foggy Tuesday morning in January 2002, the police contingent stationed outside the American Center in Kolkata was barely awake when two motorcycles drove up and unleashed a torrent of bullets on the unsuspecting policemen. Less than a mile away, other police units were practicing for a parade to be held on India's Republic Day. In an attack that lasted just 4 minutes, terrorists were able to kill four Indian policemen and injure another 18, as well as a private security guard and a pedestrian. Later that day, two groups claimed responsibility for the attack—the Harkat-ul-Jihad-i-Islami and the A. R. Commandos, later known as the Asif Reza Commando Force ARCF, (Battarcharya 2002).

Years later, some of the individuals who led this act of terror would play a role in creating a much more lethal entity, the Indian Mujahideen (IM), a terrorist group primarily based in India but also part of a wider network with operations in Bangladesh, Nepal, Pakistan, and the Persian Gulf. With financing, training, travel documents, logistics and operational support often provided by Lashkar-e-Taiba (LeT) and Pakistan's Inter-Services Intelligence agency (ISI), IM grew out of its origins into a professional terrorist group that has killed approximately 750 people in India since October 2005 and over 1,000 if going back to the January 2002 attack.

V. S. Subrahmanian et al., *Indian Mujahideen*, Terrorism, Security, and Computation,
DOI: 10.1007/978-3-319-02818-7_1,

The first terrorist attack where the name "Indian Mujahideen" emerged was a simultaneous attack on courthouses in the cities of Lucknow, Faizabad, and Varanasi in India's Uttar Pradesh state. Significantly, the group sent an email to the media minutes before the blasts, claiming responsibility for the attacks. The same email also claimed responsibility for attacks in Delhi on October 2005 and Hyderabad on August 2007 (NDTV 2007).

When these earlier attacks occurred,[1] Indian police were unaware of the emergence of IM as a group and believed the attacks to have been orchestrated by either LeT or HuJI. It was only in the aftermath of the November 2007 triple court attacks in Uttar Pradesh (UP) and IM's emailed communiqué that Indian security services began to take the emergence of a domestic Islamist terrorist group seriously (Gupta 2011).

The deadliest attack attributed to IM to date is the July 2006 Mumbai train bombings that killed 209 people and injured over 700. Mumbai's commuter train system runs along two lines through the long, relatively narrow sliver of land that is Mumbai. The line operated by Western Railways has a major hub at Churchgate at the south end of the city and another major hub in Borivli near the north end. Seven blasts rocked the city as passengers rode the trains from Churchgate. Over 2 years later, in September 2008, Indian police arrested Mohammed Israr Sadiq Sheikh for his role in the 2006 attacks; he confessed to building the bombs in a Mumbai apartment (Hafeez 2009).[2]

More recently (CNN 2013), the US State Department asserts that the Indian Mujahideen played a "facilitative role" in the infamous November 2008 attacks on Mumbai. Though the latter was primarily orchestrated by LeT with support from Pakistan's Inter Services Intelligence agency, this assertion from the US State Department is extremely important.

Indian law enforcement agencies since then have made enormous progress in their investigation and understanding of IM. A number of arrests have shed important light on IM's operations, including the arrests of Aftab Ansari and Jamaluddin Nasir (both convicted by Indian courts of carrying out the January 22, 2002 attack on the American Center in Kolkata), as well as IM leaders such as Mohammed Israr Sadiq Sheikh, the first of IM's founder arrested to date. These individuals and many others have provided valuable intelligence on IM's origins, IM's operational planning, IM's relationship with Pakistan's ISI, and IM's relationships with other terrorist groups like LeT and HuJI (Gupta 2011). As this book

---

[1] See Appendix B for a list of attacks in which IM is a leading suspect.

[2] The identity of the group that perpetrated the 2006 Mumbai train bombing is not known for certain and is a source of some debate within both Indian and U.S. security circles. Sheikh has recanted his confession to the 2006 bombings and the investigating Anti-Terrorism Squad has dropped charges that Sheikh was involved in the 2006 bombing (PTI 2013). However, the other most likely suspect, Pakistan-based LeT, helped establish IM by recruiting and training Indian operatives. It is probable that, if the 2006 attacks were primarily orchestrated by LeT, the attacks still had a substantial local component. For the purposes of this analysis, the Mumbai 2006 bombings have been considered a joint IM/LeT operation.

went to press in August 2013, IM's operational commander, Yassin Bhatkal had just been arrested at the India–Nepal border and was being interrogated by Indian law enforcement forces. It has been reported that Bhatkal has already provided information on his travel to Dubai, Pakistan, and the USA (DNA India 2013). It is likely that Bhatkal will provide further insight into IM's network of operatives and supporters as his interrogation continues.

Despite this wealth of information now starting to emerge, much in an excellent book by Gupta (2011) and work reported by Fair (2010), there have been few systematic studies of the Indian Mujahideen. Gupta (2011) provides a wealth of information on IM's origins, personnel, and operations, while Fair (2010) describes the origins of IM and its relationship to the Students Islamic Movement of India (SIMI). We treat these works as "source" data for our study, although our team has explored a vast range of other open source information, particularly from the Indian press. Moreover, as the last author served his entire career in Indian law enforcement including a stint as Director of India's Central Bureau of Investigation (CBI), our research team has had the opportunity to both look at "open source" information and draw upon the benefits of years of experience within an investigative authority.

Finally, this work blends a mix of public policy, criminology, and technology in a way that has never before been applied to analyzing IM and has only been applied to one other terrorist group, LeT (Subrahmanian et al. 2012). A brief note on our methods is therefore in order.

We use the sort of rigorous "big data" analysis that the Amazons and Ebays of the world use to model the behavior of individual consumers, models that are then used to present the consumers with product recommendations most likely to elicit a "click". Just as Amazon and EBay systematically collect consumer data, we collected data on more than 770 variables on a monthly basis from January 2002 to December 2010.[3] The data include variables describing the actions (and their intensity) taken by IM as well as variables describing the environment within which IM took those actions. We then used data mining algorithms (Subrahmanian and Ernst 2009) to identify aspects of IM's environment that predicted certain types of IM attacks one to 5 months later. These variables, in conjunction with the type of attack predicted, are represented via the methodology of temporal probabilistic rules (TP-rules), originally introduced by Dekhtyar et al. (1999) and subsequently used by Shakarian (2011) and (2012). We explain TP-rules in plain English in Chaps. 4–7; readers with no computational background should find these chapters accessible. Finally, we used the derived TP-rules and further computational methods to automatically generate policies that may help reduce IM terrorist attacks. Simplified versions of TP-rules were used to successfully model the behavior of Hezbollah, Hamas, and Lashkar-e-Taiba (Mannes 2008a, 2008b, 2011; Mannes and Subrahmanian 2009).

[3] Even though IM may or may not have existed in January 2002, actions carried out by some of its predecessors such as ARCF and elements of SIMI were coded as part of IM during the 2002–2005 time frame.

1.1 Organization of the Book

The rest of this book is organized as follows.

Chapter 2 contains a detailed description of the origins of the Indian Mujahideen (IM), IM's founders, and the groups involved in the creation of IM. It includes a description of IM's ideology, attacks carried out by IM, key IM leaders, the relationship between IM and other terrorist groups (e.g., LeT and HuJI), and the relationship between IM and Pakistani intelligence.

Chapter 3 describes the temporal probabilistic (TP) rule formalism—this chapter can be skipped by those not interested in the technology. The chapter describes how these rules were automatically extracted from our IM dataset.

Chapter 4 contains a detailed description of IM attacks on public sites (such as markets) and discusses the conditions under which IM has carried out such attacks in the past. We note for the record in this chapter (as well as in subsequent chapters analyzing IM's behavior) that past behavior is not always a good predictor of future behavior—but it is an essential starting point for any rigorous analysis.

Chapter 5 looks at a variety of bombings carried out by IM, including the bombings that were possibly carried out in conjunction with other groups such as LeT or HuJI. It identifies the conditions under which such attacks were carried out.

Chapter 6 describes the conditions under which IM carries out simultaneous and/or timed attacks in which a number of blasts occur at or around the same time, often but not always in the same city. Simultaneous attacks using improvised explosive devices (IEDs) are a hallmark of IM operations. Such attacks have been executed in Mumbai, Pune, Hyderabad, Delhi, and Uttar Pradesh among other locations. Moreover, IM has executed attacks using a variety of improvised explosive device (IED) attacks using pressure cooker bombs, bicycle bombs and car bombs, with a variety of explosives, including RDX and ammonium nitrate. Their attacks show careful planning, including methods to stage a second attack so as to target citizens escaping the aftermath of a first attack.

Chapter 7 studies the conditions that predict the total number of innocent civilians killed in IM attacks.

Chapter 8 is a technical chapter that describes how we automatically generate policies from the set of rules derived about IM. Readers not interested in the technology can skip this chapter.

Chapter 9 describes the one policy that our Policy Computation Algorithm generated to reduce attacks by IM. It discusses both what the policy says and suggests methods to implement the policies.

Chapter 10 describes the need for India to build a National Counter-terrorism Center (NCTC) similar to that created in the U.S. after 9/11. In contrast to the US, states in India wield considerable power over police and internal security, posing a significant obstacle to building an NCTC-India.

A set of appendices provides rich supplementary material. Specifically, Appendix A describes the codebook (Shakarian et al. 2012) used to code data in our IM dataset and also specifies how certain mathematical quantities used in this book were calculated. Appendix B is a list of all terrorist attacks carried out by IM and its predecessors during the 2002–2013 period. Appendix C is a list of all TP-rules presented in this book. Appendix D is a list of some of the episodes when there was a warming of diplomatic ties in the India–Pakistan relationship. Appendix E is a list of some of the IM conferences that have been held over the years; Appendix F is a list of reports of claims of responsibility issued by IM; and Appendix G is a list of reports related to IM's communications about its terror campaign.

1.2 How to Read this Book

Readers with no interest in the technology used to derive the results in this book can safely skip Chaps. 3 and 8 without losing any insights into IM's behavior. Readers with an interest in the technology will find technical details provided in these two chapters; we do not provide the detailed mathematics used to derive our algorithms here but explain it in a simple manner so as to convey the basic ideas of how these computations are done. The details are provided in the technical references provided in these chapters.

1.3 IM's Lethality: A Summary

Though IM officially announced its existence only in November 2007, many of the elements that eventually constituted IM existed for many years earlier, either in the form of the ARCF or as the SIMI. As a consequence, our study is based on data on IM attacks from 2002 through 2010; events since 2010 are also discussed but were not coded in the data collection phase.

Figure 1.1 shows the number of attacks attributed to IM and its predecessors during this time frame. Figure 1.1 shows the number of attacks, where an attack might include a series of connected incidents; that is, simultaneous bombings would be counted in Fig. 1.1 as one attack.

Figure 1.1 shows that IM does not carry out many attacks per year and that 2008, with five attacks, was the peak year for IM activity. Moreover, 2009 was a quiet year in the aftermath of the Mumbai attacks and crackdowns on IM by Indian security. After 2009, IM returned to an operational tempo comparable to that of pre-2008.

Figure 1.2 shows the total number of attacks when each event within a simultaneous/timed bombing is counted as an individual attack. For instance, in IM's 2010 attack on Jama Masjid in Delhi, there was a drive-by shooting and a car bomb. In Fig. 1.2, the drive-by and car bomb are each considered separately and counted as two attacks for that day. As another example, IM set off five bombs and

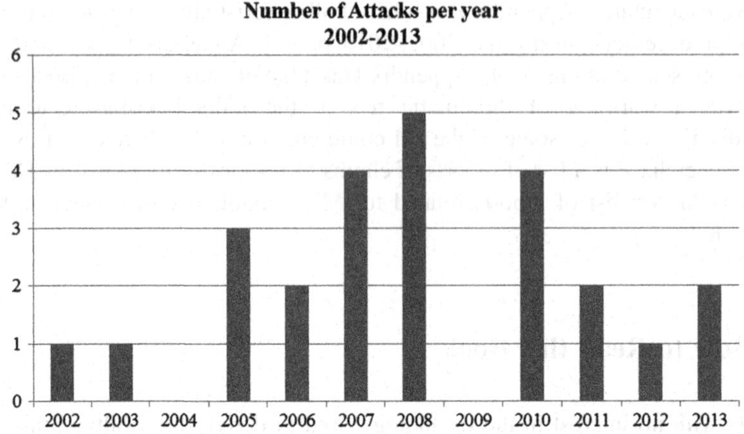

Fig. 1.1 Number of IM attacks per year, including attacks by IM's predecessors, January 2002–April 2013

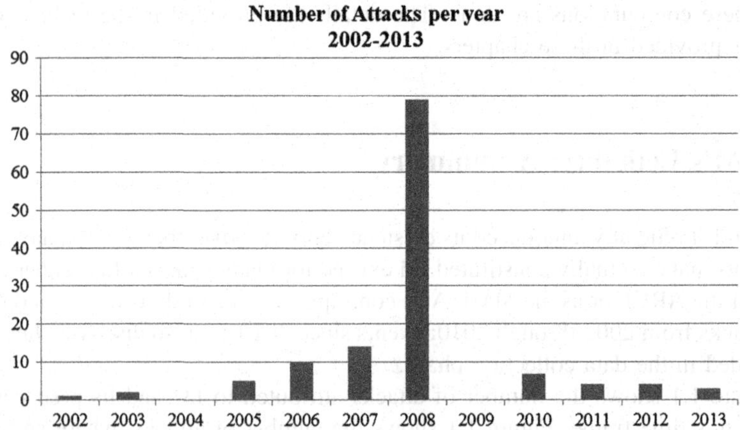

Fig. 1.2 Number of attacks carried out by IM, January 2002–April 2013, each event within a simultaneous/timed bombing considered as an individual incident

announced its presence to the world in November 2007. This attack was counted as five attacks in Fig. 1.2.

Figure 1.2 further emphasizes that 2008 was IM's most active year by far since each of the five operations IM launched that year included a large number of individual bombs. The July 2008 Ahmadabad attacks alone included dozens of bombs planted throughout the city. In contrast the two IM operations in 2011 were a single bombing in Delhi and three bombs set off in Mumbai.

Figure 1.3 shows the number killed in these attacks.

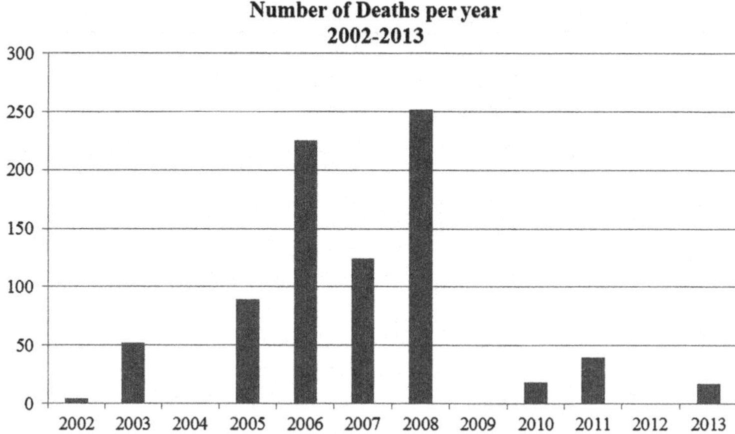

Fig. 1.3 Numbers killed in IM attacks per year, January 2002–April 2013

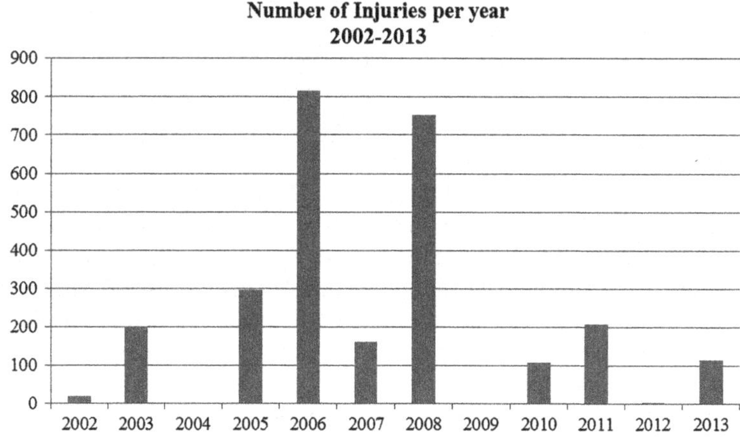

Fig. 1.4 Numbers injured in IM attacks by year, January 2002–April 2013

Figure 1.3 also highlights that IM was extremely active during the 2006–2008 period, killing over 100 people per year during that period, with the exception of 2005. During this entire timeframe, IM killed over 600 people. The Mumbai train bombings, the single deadliest operation linked to IM, in which over 200 were killed took place in 2006. After 2008, the tempo of IM operations continued, but the scale and lethality of these attacks decreased substantially. Nonetheless, Fig. 1.3 shows that despite crackdowns, IM remains a capable of launching deadly attacks.

Figure 1.4 shows the number injured during the same period.

Table 1.1 Cities where IM or affiliated/predecessor organizations have carried out attacks

Town	State	Number of attacks
Hyderabad	Andhra Pradesh	2
Barpeta Road	Assam	1
Bongaigaon	Assam	1
Guwahati	Assam	1
Kokrajhar	Assam	1
Subtotal	Assam	4
Ahmedabad	Gujarat	1
Panipat	Haryana	1
Bengaluru	Karnataka	3
Mumbai	Maharashta	3
Pune	Maharashta	2
Subtotal	Maharashtra	5
Delhi	National Capital Region	4
Jaipur	Rajasthan	1
Faizabad	Uttar Pradesh	1
Gorakhpur	Uttar Pradesh	1
Jaunpur	Uttar Pradesh	1
Lucknow	Uttar Pradesh	1
Varanasi	Uttar Pradesh	4
Subtotal	Uttar Pradesh	8
Kolkata	West Bengal	1
Total	India	30

Figure 1.4 shows that in general the years of IM's deadliest attacks (from 2005 to 2008) are also the years in which there are the most injuries in IM attacks. The attacks of 2007 appear to be an exception in which the number of injured is relatively low in proportion to the number of people killed. During the period from 2002 to April 2013, 2,679 people were injured in attacks carried out by IM.[4]

Table 1.1 below shows the locations of IM attacks that we were able to document. This table counts the number of times a city was targeted for attack, thus a multiple bombing (such as the July 2008 attack in Bengaluru in which seven bombs were detonated) in one city is one attack. The November 2007 courthouse bombing across three cities in Uttar Pradesh is counted as an attack on each of these cities.

Table 1.1 shows that IM's area of operations extends throughout the country, from Assam in the northeast to Karnataka in the southwest. Uttar Pradesh, the most frequently targeted state, is the largest Indian state by population and home to many IM operatives. The target selection also reflects IM's ideology and strategy. The two most targeted cities are India's capital Delhi and Varanasi, a holy city to Hindus. The next two most targeted cities are Mumbai and Bengaluru, India's commercial and technology capitals respectively.

---

[4] The numbers shown in Figs. 1.3 and 1.4 represent the lower bounds on casualties. For example, when it was reported that 15–20 people were injured in an incident, we counted the number of injuries as 15.

1.4 IM's Principal Areas of Operation

The Indian Mujahideen is primarily composed of two wings, one based in Northern India and the other in Southern India.

Over the years, IM's operations in Northern India grew out of the stronghold of the Azamgarh district in Uttar Pradesh. A single district with a population of over 4.6 million people, Azamgarh has a high population density and a significant Muslim population including Arabic universities. IM's operatives subscribe to the ideology of the madrassah of Dar-ul-Uloom in Deoband (also in Uttar Pradesh) often referred to as the "Deobandi" school of Islam which subscribes to a very strict interpretation of the Koran. Some IM operatives were also influenced by the blend of extremist ideology preached by Abu'la-A'la Mawdudi (Nasr 1994, 1996), a preacher who turned to political ideology including the creation of the Jamaat-e-Islami (JeI) party, which has branches in both Pakistan and India (Gupta 2011; Sikand 2006).

In addition to regions like Azamgarh and Saharanpur, IM also has significant centers of action in Deobandi madrassahs in Bharuch (in Gujarat) and Ujjain (in Madhya Pradesh).

In the north, one may also count the states of Bihar and West Bengal as areas of considerable (current or previous) IM activity.

In the south, IM has strong power bases in Maharashtra (within the urban centers of Mumbai, Pune, Surat, and Nasik), in Andhra Pradesh (where they have carried out frequent attacks in the city of Hyderabad), in Karnataka (the Bhatkal brothers who co-founded IM with others grew up in Mangalore, Karnataka) as well as in Bengaluru where numerous IM attacks have taken place. IM also has power bases in both the states of Tamil Nadu and Kerala (Fair 2010; Gupta 2011).

Figure 1.5 displays IM activity in India on a map. The background of the figure is gray, with the sea shown in blue as usual. Locations of attacks carried out by IM are shown as flame/fire icons (shown in red in the electronic version of the book)—these also include attacks by predecessor organizations of IM like ARCF as well as attacks where IM is a major suspect. Attacks usually, but not always, occur in places where a group has at least some support within the underlying population. Locations of conferences held by IM are shown via unshaded houses (with yellow borders in the electronic version). We will show later in this book that IM conferences have often preceded IM attacks, suggesting that IM conferences are used for operational planning. Shaded houses (green in the electronic version) depict locations where IM has a significant presence. Note that there are many other locations in India where IM has a presence—these icons are only placed where the presence is significant.

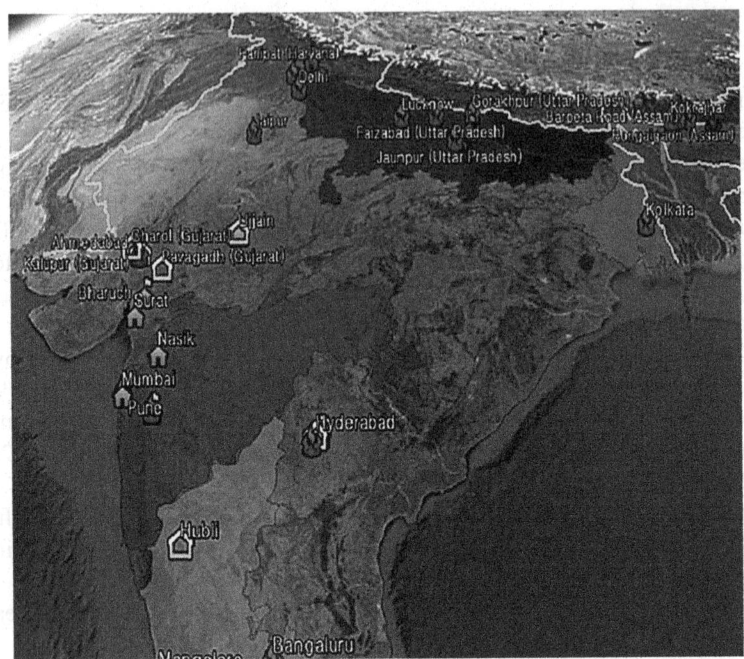

Fig. 1.5 Geographic representation of IM presence. Shaded regions show states with a documented IM presence. "Flame" icons (*red* in the electronic version) show cities/towns where an IM attack occurred. *Unshaded houses* (in *yellow* in the electronic version) show locations where an IM conference has occurred. *Shaded houses* (in *green* in the online version) show regions where an IM presence has been document. *Image* Screenshot from Google Earth. Map does not show all of India

1.5 Summary of IM Behavioral Model

Over the years, IM has carried out IED attacks on a wide variety of targets. Though most targets have been "soft" targets such as markets, IM carried out attacks on slightly "harder" targets such as several courthouses in India and the ARCF attack on the American Center in Kolkata.

Using our data-mining technology, we were able to derive TP-rules that summarize conditions under which IM launches three types of attacks.

- Attacks targeting public sites such as Zaveri Bazaar in Mumbai in July 2011;
- Bombings such as the bombing at the German Bakery in Pune in February 2010; and
- Simultaneous, timed attacks such as the nine bombs placed at seven different locations in the city of Jaipur in May 2008.

Table 1.2 Relevant variables in predicting kinds of IM attacks

	Attacks on public structures	Bombings	Consecutive or timed attacks	Total killed
Arrests of IM personnel	Yes	Yes	Yes	Yes
Communications about its terror campaign by IM	Yes	Yes	–	Yes
IM conferences	Yes	Yes	Yes	Yes
IM claims of responsibility	Yes	Yes	–	Yes
Warming of diplomatic relations between India/Pakistan	Yes	Yes	–	Yes
Internal conflict within India	–	Yes	–	Yes
Membership of IM personnel in other non-state armed groups	Yes	Yes	Yes	Yes

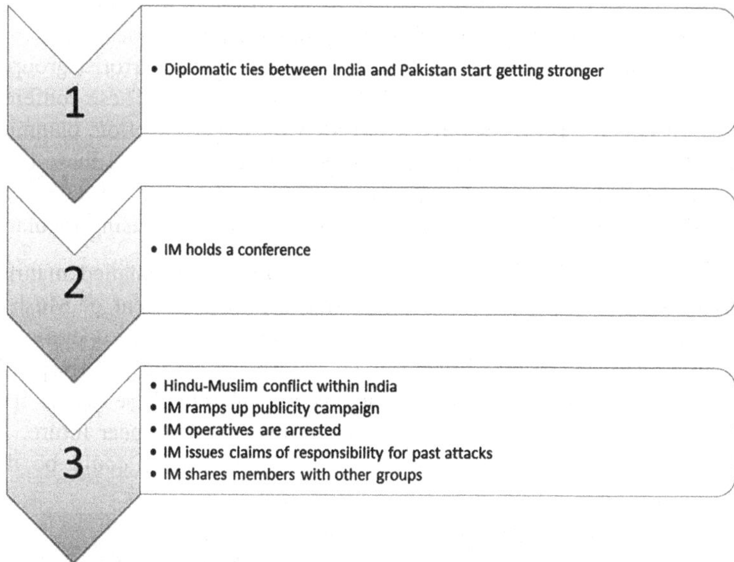

Fig. 1.6 Three steps preceding most types of IM attacks

We emphasize that though IM has carried out other kinds of attacks for which we were unable to derive rules with strong predictive power, most IM attacks to date did fall into one of these three categories.

Specifically, our findings reported in Chaps. 4–7 of this book show that the variables listed in Table 1.2 are related to the three types of attacks listed above. We note that in various cases, these variables are predictive of attacks in conjunction with other conditions.

Our findings, detailed in Chaps. 4–7 of this book, provide compelling evidence that prior to IM attacks, three things occur, not necessarily in the sequence shown in Fig. 1.6.

Figure 1.6 shows a set of steps that seem to precede every kind of attack that IM has typically launched. The progression of events seems to follow the steps listed below.

- When there is a warming of diplomatic ties between India and Pakistan, we can expect IM attacks to follow 4–5 months later. It is possible that elements of Pakistan's military apparatus feel threatened when the civilian Pakistani leadership and the Indian government come closer to signing agreements or making joint announcements of a normalization of ties between the two nations. Pakistan's military establishment derives significant benefits from stoking anti-Indian sentiment within the Pakistani population. For instance, Pakistan's military is involved in a wide variety of business activities (Siddiqa 2007) and corporations, estimated by some to exceed $20 billion per year (Al-Jazeera 2008). Were India and Pakistan to become friendly neighbors, the need for Pakistan's military would dry up, leading to an erosion of the power and money that justifies the prominence of the military in Pakistani politics.
- An IM conference often precedes attacks by 3–5 months. Terrorist groups find that most critical communication needs to be held in person. These conferences, meetings between top IM operatives, are used for target selection, planning the details of an operation, and for assigning personnel to carry out these tasks.
- Between one and three months prior to an IM attack, some subset or combination of the following four events seem to occur with surprising regularity.

  - *IM makes statements about its terror campaign.* IM's emailed manifestos, which revisited IM's grievances regarding the mistreatment of Muslims in India and Hindu violence against Muslims, often hint at future operations. It is possible that these manifestos are intended, at least partially, to inspire IM's followers. Regardless, when attacks are accompanied by these emails, it could be a sign that further attacks are coming in the relatively near future.
  - *A number of IM operatives are arrested.* These arrests could be due to increased IM activity in preparation for an upcoming attack.
  - *IM issues claims of responsibility.* IM's claims of responsibility are frequently in their emailed manifestos. When IM issues these claims, it may be a sign that future attacks are in the works.
  - *IM increasingly shares operations with other groups such as HuJI and LeT.* IM has benefitted from its links to other terrorist groups, particularly LeT (Fair 2010) and HuJI (Gupta 2011). These links allow IM operatives to gain the necessary skills and materials to carry out operations. In the past few years, the presence of several IM founders in Pakistan working closely with LeT, indicates that IM to some extent may have become a proxy of LeT. Reports of expanded links between IM and other terrorist organizations may be a signal that IM is coordinating operations or gathering materials for an upcoming attack.

Considerable early warning signs usually precede IM attacks. By monitoring the state of India–Pakistan diplomatic relations and determining whether IM is holding conferences, security officials should be able to flag whether an attack is likely or not. Should a warming of India–Pakistan diplomatic ties be followed by a conference, then there is increased reason to believe that IM attacks are likely—and if these two events are further followed by some combination of an IM publicly discussing their terror campaign, including claims of responsibility for past attacks, a number of arrests of IM operatives, and increased evidence of IM involvement with other non-state armed groups, then the likelihood of IM attacks in the near future is high.

1.6 Policies Against the Indian Mujahideen

Unlike the case of our prior research on LeT where no policies existed to reduce the different types of attacks carried out by LeT (Subrahmanian et al. 2012), in the case of IM, we were able to find one such policy.

Specifically, this policy has the actions suggested below.

1.6.1 Increased Vigilance When India–Pakistan Diplomatic Relations Warm

According to our Policy Computation Algorithm and as is clear from the discussion in Sect. 1.5 above, when there is a warming of diplomatic relations between India and Pakistan, IM carries out attacks in 4–5 months. As a consequence, when there is a thawing in India–Pakistan diplomatic ties, there is a need for increased security around sites, as well as a need for increased intelligence gathering.

Recommendation 1: Our first suggested policy action is that the state of India–Pakistan diplomatic relations be monitored very closely. This is not difficult to do, as the Government of India is party to such relations.

Recommendation 2: Our second related policy action is to dynamically monitor sentiment related to the state of Hindu–Muslim relations in India for any signs of communal tensions. This may be done in a variety of ways (under appropriate legal safeguards) by monitoring Short Messaging Service (SMS) traffic and social media sites like Twitter and Facebook as well as email traffic. For this purpose, a *Textual Intelligence System* is required that analyzes open source information as well as textual information obtained via other means.

Recommendation 3: In order to track communications between entities for evidence of upcoming religious violence, we also recommend the creation of a *Communications Intelligence System*, which is much more focused in its activities than the Government Communications Headquarters (GCHQ) in the UK. Such a system would focus on tracking—with appropriate legal authorization—phone communications, email communications, and more. These, in turn, would be passed on to the *Textual Intelligence System* described above.

1.6.2 Monitor Conferences Organized by the Indian Mujahideen

Often, IM holds conferences three to 4 months prior to an attack. As such, there is a critical need to monitor and infiltrate these conferences and to upgrade both the security threat within India and the intelligence gathering operations when IM is organizing a conference.

Recommendation 4: The Communications Intelligence System above which tracks communications between entities will help identify when and where conferences are planned. But more is needed.

Recommendation 5: In order to understand what is being said and/or discussed at IM conferences, there is an increasing need for human intelligence. In particular, IM needs to be infiltrated so that additional intelligence is available on upcoming IM operations.

Recommendation 6: In order to attend conferences within India and with trainers in Pakistan and elsewhere, IM operatives need to travel. As a consequence, a comprehensive *Travel Intelligence System* is needed in order to understand who is traveling, when they are traveling, to where they are traveling, and with whom they are traveling. This travel intelligence system needs to encompass, at the very least, the Middle East in the west through Indonesia in the east. IM, LeT, HuJI, Dawood Ibrahim's D-Company, Pakistan's ISI, and supporters and related organizations have been active across this entire region.

1.6.3 Monitor IM Rhetoric About Their Terror Campaign and Claims of Responsibility by IM

Emailed communiqués and claims of responsibility often precede IM attacks.

Monitoring communications for these communiqués can be achieved via Recommendation 2 above that should provide the textual intelligence required in order to monitor and analyze IM's communications.

Recommendation 7: As IM communiqués and claims of responsibility are often sent to the media, there should be regulations requiring the media to promptly report threats and/or claims of responsibility within a short window of time to a designated security agency, which would have the technical skills and equipment needed to evaluate the validity of the message.

Recommendation 8: IM manifestos discussing the organization's strategy and operational campaign are a warning that more attacks are in the works. When these are detected, Indian security forces should be alert, but there is a critical need to counter IM public messaging in an open manner in order to defuse a brewing situation. This can be done by counter-messaging the community that IM is seeking to influence. Counter-messaging requires relationships with moderate elements in mosques and other community organizations that are

Table 1.3 Summary of policy recommendations

Strategic policy action	Tactical implementation
Enhance vigilance and intelligence gathering after arrests of IM personnel	Improve acquisition of intelligence obtained from suspects and better aggregation of this information with other intelligence on IM activities
	Increase security at likely targets after IM arrests
	Reduce publicity given to arrests of IM personnel
Monitor communications about their terror campaign by IM	Increase security and intelligence activities when IM issues manifestos
	Develop assets and methods to counteract IM messaging intended to energize its cadres
Monitor conferences organized by IM	Increase infiltration into IM networks, intelligence gathering and electronic surveillance of IM activity
	Build a "travel intelligence" capacity to systematically monitor the movements of IM members to learn when meetings are occurring and if possible disrupt them
Monitor claims of responsibility by IM	Monitor news and social media for claims of responsibility and develop tools to identify origins of claims in real-time
	Enact legislative tools needed to require news organizations to report claims of responsibility and that state, local, and national security forces share this information in a timely manner
Monitor internal conflict within India	Develop tools to monitor Hindu-Muslim sentiment across multiple data channels including news, blogs, social media, SMS, and phone calls
	Develop security capabilities to defuse potential communal violence
Monitor reaction to warming of diplomatic relations between India and Pakistan	Increase infiltration of IM network and increased intelligence gathering and electronic surveillance of IM activity
	Develop tools to continuously monitor sentiment in news, blogs, and social media regarding India–Pakistan relations, Hindu-Muslim relations, and leadership of India and Pakistan

opposed to violence and are willing to publicly play down extremist ideology. Such "counter-messengers" need to be appropriately protected from retaliation by extremists.

1.6.4 Monitor Internal Conflict Inside India

As shown in Table 1.2 earlier in this chapter, there is a strong relationship between the state of Hindu-Muslim relations within India and subsequent IM attacks. The Babri Masjid demolition and the Gujarat riots in which Muslim communities were targeted are just two examples that are frequently cited in IM's rhetoric.

The Textual Intelligence System specified in Recommendation 2 is a method of monitoring Hindu-Muslim conflict within India. When sentiments on one or both sides are running high, one can expect subsequent IM attacks.

Thus, monitoring internal Hindu–Muslim strife within India is critical in averting subsequent IM attacks.

1.6.5 Increased Vigilance After Arrests of IM Personnel

As shown in Table 1.2, arrests of IM personnel are often followed by various kinds of terrorist attacks.

There are circumstances when IM personnel must be arrested. But extreme care should be taken in carrying out such arrests. Random security sweeps and dragnets that capture innocent civilians who happen to be inadvertently connected to some IM operatives can have the effect of enlarging the population of radicalized individuals.

Table 1.3 summarizes the set of strategic actions we recommend.

1.7 Conclusion

The Indian Mujahideen is perhaps one of the least understood terrorist groups operating in South Asia. Its origins are hazy, but there is compelling evidence that it grew out of elements associated with SIMI and ARCF. IM announced its existence only in 2007 but simultaneously claimed responsibility for attacks going back to 2005. Its founders are believed to consist of the following five individuals: Amir Reza Khan who also founded ARCF after the death of his brother Asif Reza; Mohammed Sadiq Israr Sheikh; Abdus Subhan Osman Qureshi; Riyaz Bhatkal Shahbandri; and Iqbal Bhatkal Shahbandri.

There is considerable evidence of collaboration amongst these entities that have displayed a mix of criminal activities through extortion, kidnapping rackets, terrorist attacks, and shared ideologies.

In this book, we have chosen to study IM, starting not in 2005, but going back to 2002 when one of IM's predecessor organizations, the ARCF, started launching attacks. We collected data on IM on a monthly basis from January 2002 to September 2010 and analyzed this data using our STATE data mining system. We note however that this book analyzes and discusses events up to April 2013 (even though STATE's analysis ended with the data up to 2010). Our study builds upon two previous studies on IM (Fair 2010; Gupta 2011). We also significantly leveraged work on related groups like LeT (John 2012; Tankel 2011) as well as from our own study of this group (Subrahmanian et al. 2012).

IM's, origins, goals, structure, and links to other terrorist groups and Pakistani intelligence are murky. This book cannot definitively answer those questions. Instead, it has focused on collecting truly *observable* variables about its behavior and used those observables to generate models describing conditions under which IM carries out attacks using rigorous data analytic methodologies that have been proven to work in industry. Based on these rigorous data analytic methodologies, we are also able to suggest strategies to help reduce terrorist attacks by IM.

References

Al-Jazeera (2008) Pakistani army's '$20Bn' business: Al Jazeera takes a look at how the Pakistani military is turning a profit. February 17. http://www.aljazeera.com/focus/pakistanpowerandpolitics/2007/10/2008525184515984128.htmlAccessed 22 June 2013

Battarcharya M (2002) Attack on American center in Kolkata. The Hindu. 23 January. http://hindu.com/2002/01/23/stories/2002012305090100.html

CNN (2013) India arrests Yasin Bhatkal, Indian Mujahideen terrorism suspect. 29 August. http://www.cnn.com/2013/08/29/world/asia/india-terrorist-arrest/index.html?iref=allsearch

Dekhtyar A, Dekhtyar M, Subrahmanian VS (1999) Temporal probabilistic logic programs. In Proc 1999 Intl Conf on Logic Programming

DNA India (2013) Rs 1 Lakh 'eidi' Gave Away Yasin Bhatkal's Nepal Hideout, 30 August 2013, http://www.dnaindia.com/india/1882139/report-rs1-lakh-eidi-to-wife-gave-away-yasin-bhatkal-s-nepal-hideout

Fair CC (2010) Students Islamic Movement of India and the Indian Mujahideen: an assessment. Asia policy 9

Gupta, S (2011) Indian Mujahideen: the enemy within. Hachette, Gurgaon

Hafeez M (2009) IM man's confession puts ATS in a spot. Times of India. 28 February. http://web.archive.org/web/20090303145945/http://timesofindia.indiatimes.com/711-IM-mans-confession-embarrasses-ATS/articleshow/4201005.cms

John W (2012) The caliphate's soldiers: the Lashkar-e-Tayyeba's long war. Amaryllis

Mannes A, Subrahmanian VS (2009) Calculated terror. Foreign Policy (on-line edition) 15 December 2009. http://www.foreignpolicy.com/articles/2009/12/15/calculated_terror?page=full

Mannes A, Sliva A, Subrahmanian VS (2011) A computational enabled analysis of Lashkar-e-Taiba attacks in Jammu and Kashmir. Proc 2011 IEEE European Intelligence and Security Informatics Conference

Mannes A, Sliva A, Subrahmanian VS, Wilkenfeld J (2008a) Stochastic opponent modeling agents: a case study with Hamas. Proc 2008 Intl Conf. on Computational Cultural Dynamics AAAI Press, Palo Alto

Mannes A, Michaell M, Pate A, Sliva A, Subrahmanian VS, Wilkenfeld J (2008b) Stochastic opponent modelling agents: a case study with Hezbollah. Proc 2008 First Intl Workshop on Social Computing, Behavioral Modeling and Prediction Springer Verlag, Phoenix

Nasr V (1994) The vanguard of the Islamic revolution: the Jama'at Islami of Pakistan. University of California Press, Berkeley

Nasr V (1996) Mawdudi and the making of Islamic revivalism. Oxford University Press, Oxford

NDTV (2007) Blasts in Lucknow, Varansi, Faizabad. 23 November. http://www.ndtv.com/convergence/ndtv/mumbaiterrorstrike/Story.aspx?ID=NEWEN20070033741&type=News

PTI (2013) Discharged IM co-founder to be called as defense witness. Press Trust of India 31. http://www.indianexpress.com/news/discharged-im-cofounder-to-be-called-as-defence-witness/1095565/

Shakarian J (2012) The CMOT codebook. LCCD, University of Maryland Institute for Advanced Computer Studies, University of Maryland, College Park, MD. Extended and revised by Schuetzle, B. and Nagel, M. in 2012

Shakarian P, Simari GI, Subrahmanian VS (2012) Annotated probabilistic temporal logic: approximate fixpoint implementation. ACM Transactions on Computational Logic 13

Shakarian P, Parker A, Simari G, Subrahmanian VS (2011) Annotated probabilistic temporal logic. ACM Transactions on Computational Logic 12

Siddiqa A (2007) Military Inc.: inside Pakistan's military economy.

Sikand Y (2006) The SIMI story. Countercurrents 15. http://www.countercurrents.org/comm-sikand150706.html. Accessed 7 May 2013

Subrahmanian VS, Ernst J (2009) Method and system for optimal data diagnosis, U.S. patent Nr. 7474987. 6 January 2009

Subrahmanian VS, Mannes A, Shakarian J, Sliva A, Dickerson J (2012) Computational analysis of terrorist groups: Lashkar-e-Taiba. Springer, New York

Tankel S (2011) Storming the world stage: the story of Lashkar-e-Taiba. Hurst

Chapter 2
Indian Mujahideen

Abstract This chapter traces the emergence, growth and consolidation of the Indian Mujahideen (IM). The chapter begins with a brief historical account of India's Islamist landscape and IM's emergence as an Islamic terrorist group. The chapter includes a brief threat assessment of IM examining past attacks, targets, and tactics, while detailing IM's leadership, organizational structure, and alliances with other terrorist groups operating in the region.

Since independence, India has faced many forms of extremist violence, ranging from ethno-separatist terrorism to ideological and religious extremism. However, over the past 20 years, and accelerating in the last decade, India has witnessed a new extremist threat from the combination of disaffected Indian Muslims and Pakistan-based state-sponsored Islamist terrorist groups that have infiltrated India from neighboring countries with the sole objective of fueling Islamic jihad. These threats have coalesced and are embodied in the terrorist group known as Indian Mujahideen (IM).

This chapter provides an overview of IM, beginning with the group's history as well as a discussion of the IM organization and its links to other state and non-state organizations. The chapter begins with a background summarizing IM's organizational predecessors including Jamaat-e-Islami (JeI), Students Islamic Movement of India (SIMI) and Asif Reza Commando Force (ARCF). It discusses how elements of SIMI turned to violence and how IM was established via a combination of elements from SIMI and ARCF. This will be followed by a description of IM's major operations, its evolution as an organization, and the effects of Indian counter-measures. The description of IM as an organization will include information on the group's ideology, structure, leadership, and tactical and logistical operations. Finally, this chapter will conclude with a report on IM's links with other organizations including other terrorist groups such as Harkat-ul-Jihad al-Islami (HuJI), Lashkar-e-Taiba (LeT), and Al-Qaeda, as well as IM's connections to organized crime and its state sponsorship by Pakistani intelligence.

Before proceeding, a word of caution is in order. Unlike LeT (Subrahmanian et al. 2012), which is a cohesive entity whose principal goal is to establish an

V. S. Subrahmanian et al., *Indian Mujahideen*, Terrorism, Security, and Computation,
DOI: 10.1007/978-3-319-02818-7_2,

Islamic Caliphate in the Indian subcontinent and beyond, IM is a much more disparate entity with different cells that operate more loosely. As a consequence, the study of IM poses challenges that were not present in our previous study of LeT (Subrahmanian et al. 2012).

2.1 Emergence and History of Indian Mujahideen

2.1.1 Jamaat-e-Islami, the Parent Organization

As Fig. 2.1 indicates, the organizational antecedent of SIMI and IM is Jamaat-e-Islami (JeI). A leading resource on JeI is the work of Nasr (1994). Nasr discusses how JeI, founded as an Islamist revivalist organization, has adapted to participating in Pakistani electoral politics. Nasr's biography of JeI founder Maulana Sayyid Abu'l-A'la Mawdudi (Nasr 1996) also provides valuable information.

Maulana Sayyid Abu'l-A'la Mawdudi founded JeI as an alternative to the secularizing Muslim League of Muhammad Ali Jinnah. Mawdudi (1903–1979) was born into a prominent, learned family. He received a traditional Muslim scholarly education, but became a journalist involved with politics and read extensively both of contemporary Muslim and Western thinkers. One of Mawdudi's interests was adapting modern Western ideas to an Islamic context.

Mawdudi focused on establishing an Islamic political organization. The 1920s saw an increase in Hindu-Muslim communal violence and Mawdudi was concerned about the rising Hindu nationalist identity emerging in the Congress Party.

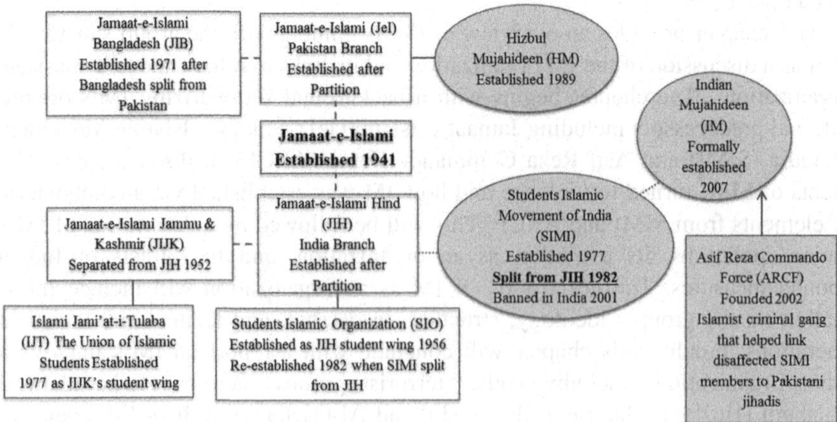

Fig. 2.1 Organizational antecedents of IM. *Bold lines* indicate formal organizational affiliation; *dashed lines* indicate organizational separations based on geography; *dotted lines* indicate ideological splits. *Shaded groups* have engaged in organizational violence, while *oval shaped* organizations have been officially designated as terrorist organizations by India. We note that though the ARCF is deemed a terrorist organization by many, it is not officially designated as such by India

Mawdudi believed that democracy in India could only work for Muslims if Muslims were the majority. However, unlike the Muslim League led by Muhammad Ali Jinnah, which Mawdudi suspected for its secular orientation, Mawdudi believed that the position of Muslims on the Indian sub-continent could only be assured through a revival of traditional Islam. As the Muslim League's proposal for two Indias (a Hindu state and a Muslim state) gained traction, Mawdudi began focusing his effort on countering Jinnah's influence in order to shape the Indian Muslim state according to Islamic law and values (Nasr 1994).

At the same time, Mawdudi opposed the worldviews of the *ulema*, the traditional Muslim religious leadership, which he viewed as too passive in the face of massive political changes. Mawdudi also rejected common Sufi practices that he felt were un-Islamic superstitions. Mawdudi formally established JeI at a conference in Lahore in August 1941 (Lerman 1981).

The author of dozens of books, Mawdudi was deeply schooled in both Islamic teaching and contemporary political discourse. While he advocated obedience to a traditional, narrow interpretation of Islamic religious law and that Muslims purge themselves of Western influences, Mawdudi also adapted Islam into a political ideology. Islamic law and practice had over the course of centuries become divorced from politics. Mawdudi sought to change that, arguing that a government run according to Sharia (Islamic law) would bring about a utopia.

Mawdudi's goal was to make Islam the supreme organizing principle for the social and political life of the Muslim community. He believed that an Islamic State could not be democratic, as all sovereign power should rest with Allah, whose representatives on earth were not a democratically elected government but rather the Islamic clergy. Mawdudi believed the supreme purpose of Islam was to establish sovereignty of Allah on earth—effectively an Islamic State. He argued that the principles and modalities for setting up God's government on earth were spelled out in the Quran and the Hadith (Lerman 1981). JeI espoused Islamic theocracy and became influential in many Muslim countries, including Pakistan, Bangladesh, West Asia, and various South East Asian countries. Mawdudi's teachings even influenced the Muslim Brotherhood ideologue Sayid Qutb in Egypt and Iranian revolutionaries (Sareen 2005).

Although Mawdudi rejected Western political ideas, his adaptation of Islam into a political ideology was influenced by his readings of Western political thought. Mawdudi rejected the Marxist idea of revolutions occurring from below and violent revolution in general. Instead, Mawdudi focused on converting the elites of society so that JeI could gradually take power. Mawdudi believed that the key to achieving his goals was in organization. Despite rejecting the Marxist strategy, he was impressed with their organizational capabilities. He once said:

> …[N]o more than 1/100,000 of Indians are Communists, and yet see how they fight to rule India; if Muslims who are one-third of India be shown the way, it will not be so difficult for them to be victorious (Nasr 1994, page 14).

Taking a page from the Communists, Mawdudi established JeI as what is known as "an organizational weapon" modeled on Lenin's conception of the

Communist party. Phillip Selznick, in his overview of the Bolshevik Party, notes that Lenin's organization was not a political party designed to translate political preferences into policy within the context of a Constitutional arrangement. The organizational weapon exists to pursue power by any means possible, including infiltrating other organizations and aligning them with the party's ideology (Selznick 1952).

To be effective, an organizational weapon requires a high degree of discipline and engagement from the party's cadre. Selznick discusses the demands made on Communist party members. Mawdudi, in recruiting and indoctrinating the cadre for JeI, made use of Muslim practice and piety—establishing a holy community of members as the basis of his organization. Although he had little use for Sufi theology and practice, Mawdudi appreciated the Sufi structure because it separated individuals from the broader society and created a rigid hierarchy that could be turned to political ends (Nasr 1994).

Mawdudi's teachings remain in effect today. Investigation has revealed that some of the top commanders of IM are deeply inspired and motivated by Mawdudi's ideology. The man may be dead, but his teachings have continued to resonate with IM terrorists to this day.

2.1.2 JeI After the Partition

After the partition of the Indian subcontinent, JeI responded to the new reality by splitting into two independent organizations, one for India and the other for Pakistan. In April 1948, at a meeting in Allahabad, the Jamaat-i-Islami Hind (JIH) was formed as the Indian branch with Maulana Abullais Nadwi as its Ameer (President). Later, other regional JeI affiliates separated from their parent organizations. These organizations were independent of one another but shared links and ideology. Each branch had to adapt its operations and objectives to the political circumstances of the area in which it was based.

Mawdudi remained head of the Pakistani wing, generally referred to as Jamaat-e-Islami (JeI), which became one of the most prominent Islamist organizations in Pakistan. It runs schools and medical facilities, participates in elections and is particularly attractive to Pakistan's educated middle class (Cohen 2004). JeI however has continued to be involved in violence. In 1979, JeI's student wing stormed the U.S. Embassy in Islamabad (Coll 2004) and, as will be discussed below, JeI was the parent organization of Hizbul Mujahideen (HM), one of the leading actors engaged in violence in Jammu and Kashmir. JeI also has links to Islamists and terrorists throughout the world. In 2003, 9/11 mastermind Khalid Sheikh Mohammed was arrested in the Rawalpindi, Pakistan home of a top JeI official (Peters 2003). In Afghanistan, Gulbuddin Hekmatyar, the brutal head of the Hezb-e-Islami group, was originally a member of the Afghanistan branch of JeI and maintained links with his counterparts in Pakistan (Haqqani 2005). In 1971, after Bangladesh became independent of Pakistan, JeI's Bangladesh branch became a separate organization Jamaat-e-Islami Bangladesh (JIB).

Because of the secular nature of independent India and the minority status of India's Muslim population, the Indian branch, Jamaat-e-Islami Hind (JIH), developed a different approach. The JIH's original objective of establishing 'Allah's government' (hukumat-e-ilahiya) was replaced with the goal of establishing religion (iqamat-e-deen). Although some scholars believe that the replacement was "more terminological than substantive-ideological" (Ahmed 2005), the realities of Indian politics and society required an extremely cautious approach that focused first on religious practice among India's Muslims before seeking broader political transformation at an indefinite future point (Ahmed 2005). JIH, which has become one of the most prominent Indian Muslim organizations, conducts charitable and educational activities within India's Muslim community and promotes religious orthodoxy. JIH also plays a prominent role as a representative of India's Muslim community, in which it focuses on the rights and identity of Indian Muslims. In defending the rights of India's Muslims, JIH has also emerged as the political order best able to preserve the rights of the Muslim minority (Lal 2004). The process by which JIH evolved from an Islamist organization that rejected India's secular democracy as *haram* (forbidden to Muslims) into a leading advocate of secular democracy is described in Irfan Ahmed's *Islamism and Democracy in India: The Transformation of Jamaat-e-Islami* (2009), which is based on 16 months of intensive fieldwork at Aligarh, a major center of JIH and its student wing the Students Islamic Organization of India (SIO). JIH has consistently rejected and condemned violence attributed to IM and SIMI as un-Islamic.

The partition and the disputed status of Jammu and Kashmir placed Jamaat-e-Islami's organization in that state in a difficult position. Jammu and Kashmir's accession to India placed the organization under the rubric of JIH, but because of Jammu and Kashmir's disputed status, the JIH leadership decided in 1952 that Jamaat-e-Islami Jammu and Kashmir (JIJK) should be a separate organization. In the intervening decades JIJK focused on building its support by running schools, providing disaster relief, and legal assistance. JIJK also participated in elections focusing on issues like corruption and the provision of government services. Although JIJK's strict interpretation of Islam was in opposition to the prevailing Sufi practices in Kashmir, over time JIJK garnered modest popular support. In 1977, JIJK established a youth wing, Islami Jami'at-i-Tulaba (IJT), which in 1979 and 1980 organized large conferences in Srinagar of international Muslim organizations (Sikand 2002).

The disputed status of Jammu and Kashmir casts a long shadow over events in the region. JIJK's constitution committed the organization to the peaceful resolution of the Kashmir situation, but talk of militancy in JIJK increased in the 1980s, although top leaders continued to publicly state their commitment to peaceful means. The increased discussion of violent resistance may have reflected the rise of a younger generation of activists through IJT (Sikand 2002).

In 1989, however, JIJK played a central role when large-scale violence broke out in Jammu and Kashmir. Initially the Jammu and Kashmir Liberation Front (JKLF) was the leading actor, but Pakistani strategists, who had their own interest in fomenting strife in Jammu and Kashmir, did not trust JKLF, since JKLF focused on an independent Kashmir, not the Pakistani goal of Kashmiri accession to Pakistan. Pakistan also believed Islamists were more reliable proxies and JIJK had a close relationship with its Pakistani counterpart (Swami 2007). On November 1, 1989, HM was formally established as an armed wing of JeI to carry out violence in Jammu and Kashmir (Rana 2006). Most of HM's cadres were recruited from JIJK's ranks and were products of its schools. Throughout much of the 1990s, HM was the most prominent of the violent groups operating there. Today, HM remains active, although other Islamist groups such as LeT have to some extent eclipsed it (Habibullah 2008).

Because of its close association with HM, JIJK was subject to massive crackdowns by Indian security forces, its offices were closed through much of the 1990s, and many of its members were arrested. While JIJK remains opposed to Indian rule over Jammu and Kashmir, JIJK's leadership in the late 1990s began to distance itself from those advocating and engaging in violence (Swami 2008).

JeI has sponsored other violent Kashmiri jihadist groups, most notably al-Badr, which was originally established by JeI to fight in Bangladesh in the early 1970s and later in Afghanistan in the 1980s (often alongside Gulbuddin Hekmatyar's Hezb-e-Islami). When JeI turned its attention to Jammu and Kashmir, al-Badr fought there, sometimes in conjunction with HM. There are reports that al-Badr split from JeI and HM over differences in negotiating with the Indian government and because HM was the preferred recipient of JeI support (Rana 2006; SATP 2013).

2.2 The Student Islamic Movement of India (SIMI)

2.2.1 The Establishment and Early Years of SIMI

SIMI's establishment in Aligarh on April 25, 1977 as the student wing of JIH was a localized aspect of an international re-invigoration of Islamism. JIH had a modest student wing, the Students' Islamic Organization (SIO), which had been founded in 1956 but was limited in scope. A younger generation of JIH activists, inspired by world events, sought a more assertive approach. The oil boom of the 1970s allowed the Gulf States to generously fund religious outreach to Muslim communities around the world. For example, the World Assembly of Muslim Youth (WAMY) was established in 1972. At the same time, the oil boom drew large numbers of Muslims from the Indian sub-continent to the Gulf for work, where they came into contact with international Muslim movements. In addition, students were taking on a more prominent role in Indian politics in general, and Indian Muslim students were inspired by the assertive activities of JeI's student wing, IJT, in neighboring Pakistan (Ahmed 2005).

Inspired by events in the world and in India, as well as their greater contact with Islamists worldwide, young JIH activists became dissatisfied with JIH's generally accommodating and moderate approach. These activists believed JIH's leadership had given up on JeI founder Mawdudi's goals of establishing the khilafat (Caliphate), an Islamist state in which "Islam [is] a 'complete program,' providing detailed instructions on all matters from the most intimately personal to collective affairs such as the state and international relations" (Sikand 2006).

According to SIMI, in the absence of the khilafat, and particularly in a pluralist, secular society, true Muslims cannot lead their lives in accordance with Islam. Thus, an organized struggle to establish the Islamic state is a solemn imperative and duty for each Muslim. SIMI embraced a more assertive approach to establishing a Muslim state in India compared to JIH's gradual "one step at a time" approach. But initially, at least officially, SIMI did not advocate violence. Jihad, which is usually defined as holy war, was defined as "learn[ing] about Islam, to act upon it and to present it to other people in the best possible way" (Siddiqi 2003). However, as will be discussed below, over time some SIMI members became radicalized and came to believe in violent Jihad and were prepared to wage war against non-believers and anyone who put hurdles in the path of the struggle for establishing a Caliphate.

Gradually, SIMI's adoption and vigorous promulgation of Quran, Jihad, and Shahadat as constitution, path and desire, respectively, branded it as a radical and violent extremist organization (Sikand 2006).

In its early years, Jihad did not translate into armed struggle. Where JIH sought to build bridges with other communities in India, SIMI sought converts to Islam. SIMI advocated on behalf of issues important to India's Muslim community but also highlighted Islam as a solution to India's problems. For example, SIMI organized an "Anti-Capitalism" week in 1983 in Kerala in order to promote Islamic economics, which, SIMI argued, would guarantee social justice (Sikand 2006).

Less than two years after SIMI's founding, international events encouraged SIMI's more confrontational and radical positions. In 1979, a relatively secular government in Iran was overthrown by Ayatollah Khomeini, galvanizing Islamists around the world in their struggle for supremacy. In the same year, the Soviet Union invaded Afghanistan, sparking a decade-long war that united Muslims worldwide in support of the Afghan mujahideen who were fighting the Soviet invaders. In addition, and particularly important to India's Muslims, General Zia ul-Haq in neighboring Pakistan had overthrown the secular government of Zulfikar Ali Bhutto and begun an "Islamization" policy. In turn, JeI became a close ally of the Zia government. This combination of events encouraged SIMI in its belief that "Islam alone was the 'solution' to the problems of not just the Muslims of India but of all Indians as such and, indeed, of the whole world" (Sikand 2006). In October 1979, SIMI held a national conference in Nagpur, attended by about 10,000 in which SIMI leaders expressed their support for the Iranian revolution and condemned the Soviet Union. The organization grew quickly (Sikand 2006).

SIMI's radical stances upset the leadership of its parent organization, JIH, which pressed SIMI to submit to its authority. Tensions between the two groups grew in 1981 when SIMI activists protested a visit to India by PLO leader Yasser Arafat. While JIH revered Arafat as the leader of the Palestinian cause, SIMI saw him as a puppet of the West (SATP 2013). In 1982, JIH reinvigorated its older student wing, the SIO, and distanced itself from SIMI (Sikand 2006). However, relations between the groups continued and according to some sources the split was not complete until 1991 when, at its fourth annual conference, SIMI's leadership embraced extremism and moved towards calls for violent jihad. Until 1991, SIMI members received ideological training from JIH teachers. This final split may have contributed to SIMI's growing radicalization, both through the loss of JIH's ideological guidance and the resignation of many of the moderate SIMI leaders (Mid-Day 2003).

Interestingly, Irfan Ahmad, who described the process by which JIH transformed into a moderate supporter of secular democracy, noted that this same process of democratization facilitated the radicalization of SIMI. First, because growing Hindu-Muslim violence undermined JIH's argument that secular democracy could best advance Muslim causes in India. Secondly, because of the spread of secular democratic values among India's Muslims, younger SIMI activists were more inclined to question traditional sources of authority within India's Muslim community and seek to interpret the religion as they saw fit (Ahmed 2009).

2.2.2 Rise in Communal Tensions

SIMI's radicalization did not occur in a vacuum. Increased communal tensions between India's Hindu majority and Muslim minority were important drivers in SIMI's evolution towards supporting violent jihad.

On December 6, 1992, a right wing Hindu mob tore down the sixteenth century Babri Masjid mosque. Located in the city of Ayodhya in Uttar Pradesh, Hindus believed that the mosque was located on the birthplace of the god Ram, the protagonist of the Hindu epic, *The Ramayana*. Many Hindus believed that the Babri Masjid mosque was built on top of a Hindu temple razed to the ground by Muslims during the Mughal period (many mosques were constructed on top of Hindu temples during the Mughal era). The Karsevaks that destroyed the mosque began to build a new temple to Ram before Indian police drove them off a few days later. Rioting throughout the country, taking over 1,000 lives, followed shortly after the destruction of the Babri Masjid mosque, poisoning Hindu-Muslim relations in India, and creating a fertile ground for a strain of radical Islam (AP 1992; Hazarkia 1992; Verhovek 1992).

The destruction of the Babri Masjid mosque was not an isolated incident. It was the climax of a period of rising communal tensions. Writing nearly two years before the razing of Babri Masjid, Indian academic Asghar Ali Engineer, who had studied communal violence in India extensively, wrote, "From October 30, 1990, the country has witnessed unprecedented communal violence. It can be said without exaggeration that after 1947 such violence has not been witnessed in the country" (Engineer 1991). He states that the Bharatiya Janata Party (BJP), a Hindu nationalist political party, in an effort to gain Hindu votes, embraced the issue of constructing a temple to Ram in the location of the mosque and aggressively campaigned for it, stoking sectarian tensions and violence. The religious disputes often exacerbated local socio-economic disputes, such as in Agra where poor Muslims and Hindu Dalits (untouchables) competed in the area's shoe-making trade. There were also Hindu extremist organizations that allied themselves to this cause and harassed and otherwise targeted Muslims, sometimes initiating outbreaks of violence. Some Muslims retaliated and radical Muslims leaders also encouraged violence. However, the majority of the dead were Muslims (Engineer 1991).

It was during this period of rising communal violence that SIMI held its fourth annual conference and took an extremist turn. In the words of a SIMI official,

> Some of our members said that the aggression of the Hindu right should be answered with aggression. We told them that we did not believe in that approach and that SIMI was not the forum for such an ideology. C. A. M. Basheer was then the all-India president of the organization. He was the one who fomented extremism in the organization (Mid-Day 2003, para 19).

According to this official most SIMI members did not support extremism and many left the organization (Mid-Day 2003). SIMI took up the Babri Masjid issue and championed it among India's Muslims, with the message that India's secular democracy could not protect them. Muslims needing to take action resonated with an increasing number of Muslims. The organization grew rapidly. In 1996, SIMI formally called for the establishment of the Caliphate (Sikand 2006).

On March 12, 1993, only weeks after the rioting in the aftermath of the Babri Masjid demolition, Mumbai was rocked by a series of a dozen blasts over two and a half hours. The bombs struck prominent commercial and symbolic sites throughout the city, including the Stock Exchange, Air India offices, top hotels, and the offices of the Hindu group Shiv Sena. Over 260 people were killed and more than 700 were injured on what became known as "Black Friday" (Gargan 1993; Spaeth 1993).

Justice BN Srikrishna (who was later appointed to India's Supreme Court) led a commission to study the causes of the 1992–1993 Mumbai communal riots and identify the responsible parties. While the commission report criticized extremist Muslim leaders and pointed out the deficiencies of the Mumbai police, it primarily laid blame on radical Hindu nationalists for instigating the violence. The Srikrishna Commision summarized the situation and explained the origin of the "Black Friday" attacks in Muslim anger, not only due to the Babri Masjid riots but because many Muslims felt that the government and police had sided with the rioters. The Commission report explained that these feelings were exploited by radical elements within India's Muslim community with support from Pakistan's ISI which, "recruited some of these angry young men by brainwashing them with the idea that they should take revenge for the humiliation and misery heaped upon them." Mumbai crime boss and smuggler Dawood Ibrahim, working with the ISI, organized these young men to carry out the March 1993 bombings in order to "engineer a fresh bout of communal riots." (Commission to inquire into the riots and incidents in Mumbai during December 1992–January 1993, 1998, Chap. 6.)

This combination of disaffected Indian Muslims, financial aid and technical support from Pakistani intelligence, and strong links to organized crime was a portent of more terror to come.

2.2.3 SIMI Radicalizes

After the destruction of the Babri Masjid mosque and ensuing violence, SIMI's rhetoric hardened. In 1996, the organization began openly calling for the re-establishment of the Caliphate—a Muslim state under Islamic law. In response to Hindu extremist rhetoric about destroying other mosques, SIMI began urging Indian Muslims to follow the path of Mahmud of Ghazni, an eleventh century Muslim conqueror who destroyed many Hindu temples (Gupta 2011). SIMI's hardline position increased its support in India's Muslim community and the more radical strains of SIMI branched out across most of India with the goal of safeguarding Islam in India. By 2001 SIMI had 400 full-time workers, known as ansars and 20,000 regular members (i.e. dedicated volunteers), was publishing periodicals, and had special wings for children and madrassa students (Ahmed 2009; Sikand 2001).

Some SIMI leaders began reaching out to Pakistani-backed extremists and terrorists, particularly their sister organization in Pakistan (i.e. the JeI). Pakistani extremist groups and the ISI sought to take advantage of turmoil in India and arranged for SIMI cadres to receive clandestine training to launch terrorist strikes that would balkanize India (Banerjee 2004).

C. A. M. Basheer, a SIMI leader who reportedly played a major role in radicalizing the organization (above) was one of the first SIMI leaders to reach out to Pakistan. Basheer fled to Saudi Arabia shortly after the Black Friday bombings in Mumbai. In the Middle East, Basheer connected with crime-lord Dawood Ibrahim who put Basheer in touch with Pakistani extremist organizations and the ISI. In the Gulf, Basheer recruited Indians for LeT and helped funnel money for terrorism to his old SIMI comrades (Gupta 2011). JeI urged Basheer to send more SIMI recruits for training in Pakistan. A member of the International Sikh Youth Federation reportedly confessed to Indian police that JeI had attempted to foster links between SIMI and Sikh extremists who, in the 1980s and 1990s, were carrying out a violent insurgency against India in pursuit of an independent Sikh homeland (Raman 2010).

2.2.4 SIMI Banned

On September 27, 2001, the government of India banned SIMI. After 9/11, SIMI praised Osama bin Laden as a hero of Islam, called the United States an enemy of Islam, and organized demonstrations in the Indian states of Madhya Pradesh, Uttar Pradesh, Maharashtra, Gujarat and Rajasthan. They also praised the Taliban for destroying the Bamiyan Buddhas and urged Muslims to "trample the infidels" (Sikand 2006).

Before the government ban, SIMI had organized two major conventions where SIMI's anti-India rhetoric was combined with a call for violence, in 1999 at Aurangabad, Maharashtra and a 2001 convention in Mumbai. Other SIMI meetings or conferences are reported to have occurred in Ujjain, Hubli, and the Charol jungle at or around the same time. SIMI's leadership made it clear in these conventions that Islam is their nation, not India, and urged Muslims to take up jihad (Swami 2008). Following the proscription of SIMI by the Indian Government and nation-wide crackdowns on its offices, the SIMI leadership went underground. The last known leaders were Shahid Badar Falahi and Safdar Nagori who served as the national president and secretary-general, respectively. While Falahi was arrested and charged with sedition and communal tension in north India in September 2001, Nagori evaded arrest until 2008.

Nagori was one of the most radical voices within SIMI, and some SIMI members hold Nagori responsible for its turn towards violence (Gupta 2011). In an April 2001 interview, Nagori described himself as "very bitter" about being an Indian and while he insisted SIMI was not fostering terror, he warned that if violence and threats against Muslims continued, "Muslims will not take it lying down" (Chakravarty 2001). Even before the ban, however, Nagori had tried to establish links with Pakistani intelligence operatives, Kashmiri militant groups (e.g. HM, LeT, and HuJI), the Palestinian group Hamas and other like-minded organizations beyond India's borders. His efforts were partly successful (Singh 2003).

2.3 Growth of Islamist Militias in India

Concurrently with the steady expansion of SIMI's radical worldview and anti-Indian activities, other Indian Muslim groups with radical extremist ideology were laying the foundation for the contemporary indigenous terrorist movement in India. Most of these organizations were formed in response to the "need to protect Islam" from the Hindu majority. During the latter part of the 1990s, Islamist extremists targeted rival Hindu rightwing groups and brought disgruntled Muslims into their fold based on real and perceived Muslim grievances. These groups had links with underworld criminal syndicates such as Dawood Ibrahim's D-Company (King 2004) and other Muslim criminal elements that sought to defend the Muslims. Most of these groups believed in the use of violence and subversion to achieve their respective goals. However, not all of them were linked to Pakistani terrorists, the ISI, or other transnational groups. Though evidence is scant, some groups, like Al-Umma and Tanzim Islahul Muslimeen (TIM), appear to be largely Indian in origin and were primarily funded by Indian Muslims engaged in criminal activities to protect their interests against Hindu rivals. Their growth in the 1990s highlights the Hindu-Muslim tensions, which fueled SIMI's turn to radicalism. Several of these groups had links to SIMI and may have provided operatives to IM. Others such as ARCF had established links with Pakistani terrorists or the ISI, links that helped enable IM in its terror campaigns. In the next several pages, we provide a short sketch of a few of these groups. This is not a complete list of Indian Islamist groups that worked with SIMI and IM as above ground front groups, recruiters, or logistical support networks.

2.3.1 Al-Umma

Al-Umma was founded in the early 1980s in the face of rising communal tensions in and around the southern state of Tamil Nadu. It emerged as a robust and organized group after the demolition of Babri Masjid (Alexander 2002; Rediff.com 1997). The group was founded by criminal elements backed by self-proclaimed Islamist ideologues like Syed Ahmad Basha who provided protection to Muslim businessmen in South India (Haque 2002). Al-Umma became popular within Islamist circles when it perpetrated bombings targeting the "Hindu power" group Rashtriya Swayamsevak Sangh (RSS) office in Chennai in August 1993. Al-Umma also masterminded the February 1998 Coimbatore bombings, in which 47 Hindus were killed and over 200 people were injured in a series of 17 blasts (BBC News 1998). The bombings were in retaliation for communal riots that took place in the city in November-December 1997. The rioting caused at least 18 deaths, mostly from the Muslim community as well as widespread destruction targeting Muslim-owned businesses. The February blasts coincided with the visit of L. K. Advani, a senior leader of the BJP, which was at the forefront of the campaign that resulted in the destruction of Babri Masjid. During the investigation of the 1998 Coimbatore blasts, it was found that Al-Umma operatives had video-taped the dead bodies of Muslims during previous communal clashes and sent the recordings to Muslim countries to mobilize sympathy and raise money for their jihad in India. The primary source of support was from the Gulf countries (Rajamohan 2005).

Al-Umma's violent metamorphosis split the organization's leadership. Many senior leaders left the organization due to the increasing radicalism while other leaders were arrested, and the state government banned the group (Subramanian 1998, 2000). While active, Al-Umma had links with, and may have received some support from SIMI (Gupta 2011). More recently, Al-Umma's name resurfaced in the recent Bengaluru blast that occurred on April 17, 2013 with the arrest of a couple of former Al-Umma operatives (Deccan Herald 2013). The blast took place near the BJP State Office located in Malleswaram area of Bengaluru. It is possible that remnants of the disbanded Al-Umma may be regrouping or joining established organizations like Indian Mujahideen.

2.3.2 Gujarat Muslim Revenge Force

As the name suggests, Gujarat Muslim Revenge Force GMRF was organized by SIMI members and LeT following the 2002 Gujarat communal riots to avenge atrocities against the Muslim community. This group centered on Ashrat Shafiq Ansari, Syed Hanif, and Hanif's wife, who together carried out the August 2003 bombings of the Gateway of India and the Zaveri Bazaar. They were arrested on August 31, 2003 and, after years of court trials, sentenced to death. Hanif had worked as an electrician in Dubai where he was recruited by LeT, possibly by C. A. M. Basheer, the former SIMI leader. This *modus operandi* of a small cell operating independently but receiving support from Pakistan raised concerns that further similar cells could be operating throughout the country and foreshadowed IM's organizational structure (Katakam 2003; Zee News 2009).

2.3.3 Asif Reza Commando Force

Many in the Indian security community believe that the Indian Mujahideen grew out of the Asif Reza Commando Force or ARCF (Indian Express 2008). ARCF is best known for a bold attack on the American Center in Kolkata in January 2002 in which five policemen were killed outside the center and 20 people were injured.

Asif Reza Khan, the Muslim Indian gangster for whom the ARCF is named, had developed links with Pakistan terrorists while serving a 5-year prison sentence in Delhi's Tihar Jail in the mid-1990s, where he met another Indian gangster with jihadi sympathies, Aftab Ansari, and the Pakistani terrorists Masood Azhar and Omar Saeed Sheikh.[1] Together the four of them began working together to raise money for jihad and to place Pakistan-trained operatives in India (Chakravarty 2002).

Ansari, Asif Reza Khan, and his brother Amir Reza Khan had active links with criminal syndicates in both the Indian subcontinent and the Persian Gulf region and operated an extortion and abduction network in India. For instance, in July 2001, Ansari, Mohammed Sadiq Israr Sheikh, and the Khan brothers kidnapped a Kolkata businessman, Partho Burman. Mohammed Sadiq Israr Sheikh is also believed to have been one of the four attackers involved in the American Center attack.

The Khan brothers, along with Mohammed Israr Sadiq Sheikh, had had working ties with the Pakistan based terrorist groups JeM and LeT, as well as the HuJI terrorist group in Bangladesh. Ansari and Amir Reza Khan had plotted many

[1] Masood Azhar is the founder of the Pakistani terrorist group Jaish-e-Muhammed (JeM), which is generally held responsible for the December 2001 attack on India's parliament. Azhar and Sheikh were released in the December 1999 hostages-for-prisoners deal when Pakistani terrorists hijacked Indian Airlines flight IC-814 (Jain 2013). Omar Saeed Sheikh is a militant of British and Pakistan origin, best known for the abduction and murder of *Wall Street Journal* reporter Daniel Pearl. He had links with a number of terrorist groups in Pakistan and beyond, including JeM. He was sentenced to death in Pakistan but is appealing the sentence (Imtiaz 2012).

criminal acts and were responsible for arms trafficking including weapons seized in November 2001 in the Patan district (Rediff.com 2002).

Two SIMI figures who would go on to be central to the emergence of IM, Mohammed Sadiq Ishrar Sheikh and Riyaz Ismail Shahbandri (better known as Riyaz Bhatkal), separately made contact with Ansari and Khan, who facilitated their travel to Pakistan for training. Khan (based in India) and Ansari (based in Dubai) urged Sheikh and Bhatkal to recruit a network of disaffected Indian Muslims and then helped arrange their travel to Pakistan for training as well as to facilitate the flow of money and explosives to India for operations. This helped lay the seeds for the network that formed IM (Gupta 2011).

Asif Reza Khan was arrested again in New Delhi on October 29, 2001, along with a Pakistani accomplice, Arshad Khan. The Gujarat police killed Asif Reza Khan during an escape attempt in December 2001. Khan's brother Amir Reza Khan took over the organization and established ARCF which perpetrated a revenge attack on the American Center in Kolkata (USIS) on January 2002 (SATP 2013).

Aftab Ansari, the Dubai based senior operative of ARCF, claimed responsibility for this attack. Aftab Ansari was arrested in 2002 and is currently jailed in India for the American Center attack. India's National Investigation Agency (NIA) believes Amir Reza Khan is hiding in Karachi, running an international kidnapping and extortion operation, and funding IM's activities in India. NIA has announced a reward of 10 lakh Indian rupees (approximately $20,000) for his arrest (Tiwary 2012).

Indian investigators believe that the founding members of the Indian Mujahideen are:

- Amir Reza Khan of Bihar;
- Mohammed Sadiq Israr Sheikh of Azamgarh of Uttar Pradesh;
- Abdus Subhan Usman Qureshi, originally from Rampur in Uttar Pradesh but grew up in Mumbai;
- Riyaz Ahmad Mohammed Ismail Shahbandri, better known as Riyaz Bhatkal, originally from Mangalore in Karnataka but grew up in Mumbai; and
- Iqbal Shahbandri, better known as Iqbal Bhatkal, the brother of Riyaz Bhatkal.

Because of the overlap between IM's current leadership and the ARCF, as well as the fact that Mohammed Israr Sadiq Sheikh was already connected to both SIMI and ARCF prior to the formation of IM, many in the Indian security establishment believe that IM grew out of both SIMI and the ARCF. However, we emphasize that the precise origin of IM is still uncertain.

2.3.4 The Hyderabad Connection

The city of Hyderabad was a particular nexus of Islamist groups that linked with SIMI and IM and provided an infiltration opportunity for Pakistani extremists. In 1985, after a round of communal violence, Abdul Karim Tunda and Azam Ghouri started TIM, a Muslim defense group. But, along with Dr. Jalees Ansari, they also were working for Azam Cheema, LeT's director of Indian operations. On the first anniversary of the destruction of Babri Masjid, Ansari set off a series of blasts in Mumbai and Hyderabad. Ansari was arrested while Tunda and Ghouri fled to Pakistan. Ghouri received training from LeT while in Pakistan and Tunda helped coordinate LeT networks in India, arranging funding for cells and travel for recruits. Several years later, Ghouri returned to India; in 1999 he attended a SIMI meeting in Hyderabad and started the Indian Muslim Mohammadi Mujahedin, which carried out a series of attacks on Hindus who were believed to have humiliated Muslims. On February 6, 2000, LeT leaders mentioned their new wing in Hyderabad and its leader in a speech, drawing the attention of Indian security forces. Ghouri was killed when Indian police attempted to arrest him on April 6, 2000. Tunda, working for LeT, also planted a Pakistani-born operative, Saleem Junaid, into Hyderabad. In the late 1990s, Saleem Junaid began establishing a cell in Andhra Pradesh. It is reported that he was also offered the leadership of the Andhra Pradesh SIMI chapter (Swami 2000).

Hyderabad continued to be a center for Islamist activity. SIMI worked closely with the Hyderabad-based Tehreek Tahaffuz-e-Shair-e-Islam and the radical Islamic vigilante group, the Darsgah Jihad-o-Shahadat (Institute for Holy War and Martyrdom), which recruited for SIMI in southern India (Government Report 2001; Gupta 2011; Sharma 2008).

2.4 Indian Mujahideen: A Jihadi Hybrid

2.4.1 The Gujarat Riots and Establishment of Indian Mujahideen

On the morning of February 27, 2002, the Sabarmati Express, filled with Hindu pilgrims returning from Ayodhya, pulled into Godhra station. The pilgrims were returning from a visit to the site of the Babri Masjid mosque where they called for constructing a temple to the Hindu god Ram. Godhra in Gujarat is a predominantly Muslim area and there was a confrontation between the Hindu passengers and the Muslim vendors at the train station. The train was set on fire and 59 passengers, Hindu pilgrims, including 27 women and 10 children, were killed. The state of Gujarat erupted into riots in which about 1100 people were killed. About two-thirds of the dead were Muslim (Nanavati 2008). Throughout India, many Muslims saw these riots as a repetition of the events of 1992, in which the state at best failed to intervene and at worst provided assistance to well-organized Hindu extremists who carried out attacks on Muslims (Gupta 2011).

For SIMI, in particular, the Gujarat riots contributed to their growing extremism. The riots re-affirmed the belief of SIMI's radicals that they must defend the Muslim community and avenge these attacks. The violence done to the Muslim community in the Gujarat riots has been a frequent theme in IM manifestos (see below). But SIMI leaders were also infuriated by their organization's ban, which they felt reflected the government's double standard on terrorism, which cracked down on Muslims while allowing Hindu extremists to operate openly and perpetrate violence (Sikand 2006).

SIMI leaders continued to meet and try to continue the organization's work under Nagori's guidance. They attempted to establish front organizations to raise money, distribute literature, and stay in touch with their cadres. Although these meetings were frequently in remote locations and included physical training and combat, SIMI's leaders remained ambivalent about turning to full-scale violence (Gupta 2011). Other groups of SIMI members openly disavowed violence and sought to rehabilitate the organization (Swami 2008).

Some SIMI members did not share this restraint. There were several attacks or attempted attacks carried out by groups linked to SIMI in the early part of the 2000s (see the Attack Table for a few examples). IM coalesced around a group of SIMI activists who had studied and debated together in the late 1990s in Trombay. Their discussions focused on Hindu-Muslim relations and the destruction of Babri Masjid. The group included Mohammed Sadiq Israr Sheikh, Abdus Subhan Qureshi, Tariq Ismail, and the brothers Riyaz Ismail Shahbandri and Iqbal Ismail Shahbandri (better known as Riyaz and Iqbal Bhatkal for their hometown of Bhatkal). Several of these individuals had moved towards violent jihad in the intervening years and Riyaz Bhatkal and Israr Sheikh were linked through ARCF (Gupta 2011). Reportedly in the summer of 2004, the Shahbandri brothers organized a meeting at Jolly Beach in their home village of Bhatkal in Karnataka. The attendees included their old SIMI contacts that collectively launched IM (Swami 2010). However, we note that Indian investigators also believe that Amir Reza Khan of the ARCF was also a founder of IM though we were not able to document his presence at this meeting.

The first attack generally attributed to the newly established IM network was the February 2005 bombing of Varanasi (a holy city for Hindus). The Varanasi attack was followed by several other increasingly deadly attacks that may have been IM operations. At this point IM was not claiming responsibility for attacks (the Varanasi attack was initially believed to be an accident). Indian authorities, unaware of the existence of IM, believed that LeT or another Pakistan-based group was responsible (Fair 2010). Most of the pre-2007 attacks by IM, including the devastating July 2006 Mumbai train bombings that took over 200 lives, were attributed to IM retroactively after the capture and interrogation of various IM operatives (Gupta 2011).[2]

[2] As discussed in Chap. 1, it is unclear which terrorist group was responsible for the 2006 Mumbai train bombing. The primary source for the claim IM carried out the attack, Mohammed Sadiq Israr Sheikh, has recanted his statement claiming that it was obtained under torture (Jaleel 2013). However the other primary suspect is LeT, which is closely linked to IM and almost certainly worked with its Indian supporters, who ultimately coalesced into IM, to carry-out the attack. Therefore in this study, the 2006 Mumbai train bombing is considered a joint IM/LeT attack.

2.4.2 Public Emergence of Indian Mujahideen

While it is difficult to establish the actual date when IM was formalized as a group, the name emerged for the first time when the group claimed responsibility by sending an email to major media outlets *moments before* near-simultaneous terrorist attacks on the courts in three major cities in the north Indian state of Uttar Pradesh (Varanasi, Faizabad and Lucknow) on November 23, 2007 (CNN/IBN News 2007). The email began with Quranic verse urging Muslims to fight the impious and claimed to have carried out the "Islamic raids" in the name of Indian Mujahideen. The email explained that it was targeting the courts because the police had arrested innocent individuals for terror attacks and because local attorneys had attacked Pakistanis who were in custody for plotting attacks on Rahul Gandhi (scion of the Nehru-Gandhi clan that had dominated Indian politics since independence). The email went on to state that the attacks were also in retaliation for Hindu attacks on Muslims in 1992–1993 and in 2002. Finally, the email insisted that IM was not linked to Pakistani terrorist groups or intelligence agencies (Gupta 2011).

Indian police initially discounted this new group, assuming it was a false front intended to throw off the investigation and focused on HuJI and arrested several suspected HuJI operatives (Gupta 2011).

Starting in May 2008, IM carried out a string of urban terrorist attacks across India, taking over 160 lives, leaving little doubt of the emergence of this new terrorist threat. Several of these attacks were also accompanied by an emailed manifesto issued just before or during the attack. The email, sent moments before the bombings in Ahmedabad took over 50 lives, gave a list of rationales for IM's attacks and threatened Hindu leaders who incited violence against Muslims, police for standing aside during attacks on Muslims, and Muslims who spied on their community for the government. The email addressed to Indian state governments stated:

> You agitated our sentiments and disturbed us by arresting, imprisoning, and torturing our brother in the name of SIMI and the other outfits… We hereby notify you… to release them all, lest you become our next targets… (Outlook India 2008, para. 5).

The email cited past operations such as the July 11, 2006 Mumbai train attacks as an example of the consequences if the Indian government did not meet IM demands, including the end of attacks on Muslims by police and Hindu mobs. The communiqué also warned the lawyers of Uttar Pradesh that the courthouse bombings would be repeated because they had failed to defend IM's "Muslim brethren." Finally, the email threatened media outlets for their "propaganda war against Muslims" and claimed that IM was a solely Indian organization and the attacks were solely their responsibility, asking LeT not to claim responsibility (Outlook India 2008).

In March 2008, Safdar Nagori, the General Secretary of SIMI was arrested. According to some reports, at this point, the remaining SIMI leaders, Abdul Subhan Qureshi and Qayamuddin Kapadia, threw their lot in with the radicals of IM, giving further impetus to the string of terror attacks starting in May 2008 (Gupta 2011).

2.4.3 Crackdown and Continuing Operations

The India-wide serial blasts of 2008 and the November 26–28 Mumbai attack carried out by LeT led to widespread crackdowns and arrests that disrupted IM's pan-India network. Indian police achieved a major breakthrough in unravelling IM's nascent Jihadi network on September 19, 2008, less than a week after IM serial bombings struck New Delhi. During a search and sweep operation in Jamia Nagar, New Delhi police came upon an IM safe house. In what became subsequently known as the Batla house encounter, police killed two Indian Mujahideen leaders, Atif Amin and Muhammad Sajid, while Mohammed Saif was arrested. According to the Hindustan Times (2013), the Batla house encounter, in which a police officer was also killed, was not pre-planned but rather occurred when police raided the location based on a tip-off. Combined with the arrest of car-thief Afzal Mutalib Usmani who worked with IM, police gained valuable intelligence on IM and arrested dozens of IM operatives, rolling up IM cells throughout the country (Gupta 2011).

Most of IM's top leaders, including the Bhatkal brothers (Riyaz and Iqbal), Amir Reza Khan, and Abdul Subhan Qureshi escaped arrest. The first three are currently believed to be in Pakistan (though as this book went to press, Yasin Bhatkal, IM's operational commander, was captured in an Indian operation on the India-Nepal border), continuing to coordinate IM operations, while Qureshi's whereabouts are unknown (Gupta 2011). There are separate reports stating that Iqbal Bhatkal may be in the tiny emirate of Sharjah in the UAE.

Though IM's network had been damaged in the wake of the Batla House encounter, it remained capable of launching terror attacks. After a hiatus of almost 14 months IM struck the popular "German Bakery" in the western Indian city of Pune. This represented a new type of target for IM because it was a location frequented by both Indians and foreigners (Times of India 2010).

The Pune attack may have been an early operation of the Karachi Project where both LeT and IM, with support from the ISI, conspired to launch terror attacks in India. Under U.S. pressure, Pakistan was forced to crackdown on groups carrying out attacks in Jammu and Kashmir in 2003. LeT sought to not only continue attacks on India, but also to build the capacity to do so through Indian front groups. Indian Muslims are recruited and sent to Pakistan, via Bangladesh, Nepal, or the Gulf, using legitimate Pakistani passports for travel once they have left India on their Indian passports. It is believed that the Pakistani passports are issued through the ISI. In Pakistan they receive training and then return to India to carry out attacks, while receiving direction from Pakistan. It is believed that the Bhatkal brothers and Amir Reza Khan are coordinating these operations from Pakistan. An important advantage of this arrangement is that it allows LeT and its ISI sponsors to continue to strike India but with plausible deniability by operating through IM (Roul 2010).

It is possible that LeT's David Coleman Headley, who is currently serving a 35 year sentence in the U.S. for scouting out possible attack targets in the 2008 Mumbai attacks, also scouted targets in Pune (such as the German Bakery) during a reported visit to the Osho Rajneesh spiritual center. This would highlight the growing coordination between IM and LeT, institutionalized in the Karachi Project (Roul 2010).

Shortly after the Pune attack, IM operatives attempted another major attack in Mumbai. However, the Mumbai police's Anti-Terrorism Squad (ATS) foiled this attack by arresting two IM terrorists identified as Abdul Latif and Riyaz Ali, who were allegedly planning to attack the headquarters of the Indian oil company ONGC, along with the bustling Mangaldas Market and Borivili's Thakkar Mall in Mumbai (NDTV News 2010a).

In June 2010, the Indian government outlawed the Indian Mujahideen and all its allied groups and front organizations under the Unlawful Activities (Prevention) Act, 1967. Highlighting its growing profile as a major terrorist organization, IM was also designated by the U.S. State department as a Foreign Terrorist Organization in September 2011 (U.S. Department of State 2011). In 2012 the UK proscribed IM under the Terrorism Act 2000 (BBC News 2012).

SIMI leaders had challenged the government's ban of their organization, arguing that they expelled any members with links to terrorism, but in February 2012, the ban was extended for a further 2 years (PTI 2012).

IM has continued to carry out attacks throughout India, often striking cities it has struck in previously such as Varanasi, Bengaluru, and Pune. While the attacks have not been as frequent or effective as those from 2006 to 2008, IM remains a dangerous organization. IM's deadliest strike since its serial bombing campaign of 2008 came in July 2011 when three synchronised blasts at prominent locations shook India's commercial capital, Mumbai. There was no claim of responsibility, but investigations found the involvement of a Bihar-based terror module, which received aid from a Pakistani operative. Yassin Bhatkal (given name Mohammed Ahmed Sidibapa), the current leader of IM in India, masterminded the operation (Times of India 2012a).

In August 2013 (after this book was already in press) Indian security forces arrested Yasin Bhatkal, along with another IM operative, on the Indian-Nepalese border (NDTV 2013). Interrogation of Yasin Bhatkal will lead to further revelations into the role of IM in the 2006 Mumbai train bombings, the precise nature of the support IM receives from LeT and Pakistan's Inter-Services Intelligence agency, as well as the links between IM and other armed groups such as HuJI. Initial reports (DNA India 2013) indicate that Indian security officials monitoring Bhatkal's family identified him through a wire transfer to his wife. The same report goes on to state that Bhatkal travelled to Dubai, Pakistan, Nepal, and the USA during this time.

2.5 Organizational Overview

2.5.1 Structure and Finance

Overall, IM is not a hierarchical terrorist group like LeT, but rather a network consisting of small cells of individuals with fluid links to IM's leadership, other cells, local Islamist groups, and criminal gangs (Swami 2010).

However, intelligence gleaned from arrests (Gaikwad 2009) in the wake of the 2008 Batla House encounter indicated that IM has established several internal units to carry out special operations.

- The Shahabuddin Ghouri Brigade oversees attacks in southern India and is reportedly headed by Amir Reza Khan who is based in Pakistan.
- The Muhammad Ghaznavi wing oversees attacks in northern India.
- The al-Zarqawi brigade targets important individuals.

IM also has a robust media wing responsible for issuing the emailed communiqués under Mansoor Peerbhoy, which is headquartered in Pune, Maharashtra (Hafeez and Hafeez 2009).

IM cells also set up front groups to facilitate their covert activities. The 2009 arrest of Tadiyandavede Nasir, the lead suspect in the 2008 Bengaluru bombings revealed that he and other IM leaders were operating under the guise of a Sufi sect known as Noorisa Tariqat, which has branches in many parts of southern India, including Kerala and Andhra Pradesh. The investigation also showed how IM cells worked semi-independently, but leaders met fairly regularly to exchange critical supplies such as bomb-making materials and also to compare notes on operations (Johnson 2009).

In the wake of the 2008 crackdown, many of IM's top leaders are still absconding and believed to be in Pakistan and the UAE. IM's command structure has changed to reflect this situation. Recent media reports state that IM has been morphed into many new offshoots such as the Bullet 313 Brigade and the Jama't Ansar-ul Muslimeen (Rediff News 2011; Sharma 2013a). It appears these offshoots act as covers for IM and are likely to be involved in recruiting and fund raising activities in Uttar Pradesh, Andhra Pradesh, Kerala and Karnataka.

IM's primary source of funding appears to be South Asians in the Persian Gulf and Muslim criminal gangs such as Dawood Ibrahim's D-Company. Both sources of funding use the informal hawala networks to channel money from donors to recipients. Reportedly Muzaffar Kola, a native of Bhatkal (also home to the Bhatkal brothers who now run IM), and a member of Dawood Ibrahim's criminal network, has funneled about 5 crore rupees (about $1 million) to support IM operations (Nanjappa 2012).

IM has also attempted to raise money through kidnapping and extortion rackets. For instance, three cases have been registered against Amir Reza Khan in Kolkata for making threating calls to Kolkata businessmen and demanding huge sums of money. Moreover, in 2001, Asif Reza Khan and Amir Reza Khan were involved in the kidnapping of yet another Kolkata businessman, Partho Burman. More recently, IM has been involved in an extortion and attempted murder of a businessman in the Kurla region of Mumbai in connection with the criminal gang of Fazlur Rehman. According to the Times of India (2012c), Fazlur Rehman's gang has been involved in various kidnappings, though convictions have been hard to secure due to witnesses recanting their stories.

2.5.2 *Areas of Operation*

IM cadres have been arrested from different locations, demonstrating the geographical spread of a terror network that now spans the length and breadth of India. IM is reportedly active in Bihar, Delhi, Uttar Pradesh, Gujarat, West Bengal, Maharashtra (Mumbai and Pune in particular), Karnataka as well as Kerala.

Though support for fundamentalism is common in some Muslim neighborhoods, support for Islamic jihad is limited and not rampant in urban centers. Islamic fundamentalism is strongest in some pockets of southern India (such as Kerala and Karnataka) and northeastern India (particularly areas bordering Bangladesh and Myanmar), in the villages of Uttar Pradesh, and in Jammu and Kashmir. Rajasthan and Orissa are hibernation grounds for Islamic militants and they often escape to these places after perpetrating attacks.

IM does not yet have operations outside of India. But its operatives and recruits take multiple paths to reach Pakistan for training, most notably through Bangladesh, Nepal, and the Persian Gulf countries (Gupta 2011).

2.5.3 *Membership*

IM members are mostly Indian though a number of foreign operatives (from Pakistan and Bangladesh) have been arrested in conjunction with IM attacks—and there is evidence of other foreign nationals from Gulf States funding IM. But beyond that, their backgrounds are varied, with recruits coming from every part of India, and from diverse economic and educational backgrounds. Many come from the middle classes and have a good education. For instance, one of IM's top leaders, Abdus Subhan Usman Qureshi received considerable computer training in Mumbai and worked for two computer firms. Other IM figures had links to India's criminal underworld. Even in terms of religious practices, IM recruits were inconsistent. Many had only limited religious education and were not terribly religious in their personal behavior. In fact, a few IM operatives confessed to police that their real ambition was to see their name on *India's Most Wanted*. The consistent theme in the motivations of IM members was seeking revenge for Hindu attacks on Muslims, particularly the violence surrounding the destruction of Babri Masjid and the riots in Gujarat a decade later (Gupta 2011).

2.5.4 *Tactics and Training*

IM's primary *modus operandi* has been IEDs (including pressure cooker bombs, bicycle bombs, and car bombs), although in a few cases, they have also carried out attacks with firearms. They have not yet carried out a fedayeen or suicide bomb attack. IM's most devastating attacks have involved setting off a large number of bombs in crowded, public places, nearly simultaneously. In the Ahmedabad bombing, IM carefully sequenced the detonations to maximize casualties—a series of smaller bombs detonated throughout the city and then a car bomb exploded near the hospital once a crowd had gathered. Although a substantial number of IM's homemade bombs do not detonate effectively, the ability of the organization to construct a large number of bombs (often from raw ingredients such as ammonium nitrate), plant the bombs without being detected, and detonate the devices almost simultaneously indicates a substantial degree of technical skill.

Another area where IM has shown at least some technical skills is in its emailed manifestos, which are also carefully coordinated with the attack. IM has been able to hack Wi-Fi accounts and use other means to make it harder for investigators to follow a cyber-trail (Gupta 2011). Indian investigators also say that IM operatives are required to leave their mobile phones behind during periods when they are engaged in an attack.

There were a number of IM training camps held in various forest locations in South India during 2007 and 2008 where the presence of former SIMI cadres and Pakistani origin nationals were reported. According to media reports, between April 2007 and December 2007, at least six training camps were held at Castle Rock and Dharwad in Karnataka, at Charol near Indore (Madhya Pradesh), and in the Nagaman jungles (Kerala). In early 2008, there was a camp in Pawagarh jungles near Vadodra (Gujarat). While this training was important, however, the most capable IM operatives also trained in Pakistan (Gupta 2011).

2.6 IM's Links to Other Organizations

2.6.1 IM and SIMI

Members of IM are mainly drawn groups such as the Students Islamic Movement of India (SIMI) and terrorist groups such as ARCF in India, LeT in Pakistan, and the Bangladesh-based HuJI. In particular, SIMI has been a major source of recruits for IM. But the exact nature of the relationship remains an open question. There is some evidence that IM is merely a new iteration of SIMI. This view is supported by the testimony of captured IM/LeT operative Zabiuddin Ansari (aka Abu Jundal), who was present in LeT's control room during the November 2008 Mumbai attacks. Ansari told Indian police that SIMI was the "backbone" for Islamist groups in India and had been involved in every major terror attack, providing local support and reconnaissance (Sharma 2012). Other analysts note that several SIMI members have come forward and surrendered to Indian police in an effort to regain the organization's legitimacy and new SIMI leaders have rejected IM's terrorism as hurting their cause (Fair 2010). At the same time, a number of Indian security officials believe that IM"s origins are rooted as much in the ARCF as in SIMI.

2.6.2 IM and Pakistan

There is now strong evidence that IM is supported by both Pakistan's ISI and LeT (Hindustan Times 2013; Sharma 2013b; Singh 2012). Many view LeT itself to be a creation of the ISI and a proxy force for the ISI (Reidel 2013). While IM may be a product of Muslim dissatisfaction with the situation within India, Pakistani intelligence and terrorist groups are exploiting this dissatisfaction to sow dissension and carry out terrorist attacks within India. But it is also possible, as the Maharashta anti-terrorism squad stated in its charge-sheet for the July 13, 2011 serial bombings in Mumbai, "The IM has been expressly created by (the) ISI of Pakistan ostensibly to spread terror in this country through Indian front outfits" (Times of India 2012b). Several other sources also argue that IM was effectively floated by the ISI when its operations in Jammu and Kashmir were limited under U.S. pressure (Shahzad 2003), as part of the "Karachi Project" (Shashikumar 2008), or as the joint program of several Pakistani terrorist groups (Rediff.com 2008b).

However, even if IM was primarily an Indian domestic phenomenon, there is little question that it "became far more lethal than it otherwise would have been without external support" (Tankel 2011).

Pakistan provides a safe haven and extensive expertise for training. These features are essential for a terrorist group to become effective, as IM undoubtedly has. Pakistani intelligence and terrorist groups have facilitated travel so IM recruits can reach Pakistan and become more effective operatives. Finally, Pakistani intelligence and terrorist groups have provided critical supplies, particularly explosives but also forged identity papers to IM operatives.

Below are discussions of a few cases of Pakistani training of IM personnel.

One of IM's top leaders, Mohammed Sadiq Israr Sheikh, from Azamgarh in Uttar Pradesh, is believed to have traveled from India to Bangladesh in 2000 on a legitimate Indian passport, and from there to Pakistan on a genuine Pakistani passport arranged by the ISI. In Pakistan, he met with LeT commander Azeem Cheema in Bahawalpur and trained at an LeT training camp near Muzaffarabad. At least another 10 people from Azamgarh alone who traveled to Pakistan through various intermediate countries have been identified.

In February 2012, Haroon Naik was arrested by the Maharashtra Anti-Terrorism Squad in connection with the July 13, 2011 triple blast case in Mumbai and is reported to have told investigators that he had met not only LeT operations chief Zaki-ur-Rehman Lakhvi in Pakistan, but had also attended an inspirational lecture by Osama bin Laden just a month prior to the 9/11 attacks. He was trained at LeT camps in Pakistan and also in terrorist camps in Afghanistan (Hindu 2012).

The case of Babu Bhai, an Indian criminal recruited by HuJI, is also instructive. After training in Pakistan (his travel was facilitated by the crime-boss Aftab Ansari), Bhai returned to India where he recruited Indian Muslims for training in Pakistan. He helped his recruits travel via Bangladesh, while also smuggling the explosive RDX, provided by HuJI, into India. Between 2004 and 2007, this channel provided the explosives for many of IM's first attacks (Gupta 2011).

A more recent example of how Pakistani support facilitates IM operations is the case of Zabiuddin Ansari (aka Abu Jundal), the IM-LeT operative who, when he was arrested in Saudi Arabia, was traveling with a Pakistani passport. The travel documents allowed him to visit Saudi Arabia to recruit Indians for LeT/IM operations, meet donors for fund-raising, and meet Indian-based IM leaders to coordinate upcoming operations (Biswas 2012).

2.7 Conclusion

The Indian Mujahideen has carried out numerous attacks in India. Though its origins remain shrouded in some mystery, there is clear evidence that it grew out of elements of both ARCF and SIMI. There are also many who believe that IM is merely an Indian arm of the Pakistani terrorist group Lashkar-e-Taiba (John 2012; Subrahmanian et al. 2012).

With financial, logistical, education and training support from Pakistan and the ISI, IM has grown into a significant threat within India—a kind of Pakistani fifth column with the goal of destabilizing India.

Moreover, unlike the November 26, 2008 Mumbai attackers, IM operatives are intimately familiar with their locales in India and have shown the ability to carry out simultaneous attacks in multiple locations across India—but noticeably in areas where they know the region well—such as Mumbai, Pune, Hyderabad, Karnataka, and Uttar Pradesh. Not surprisingly, many of IM's top leaders were either born or brought up in these regions.

Last but not least, IM has operatives who are well educated and well trained. Many of their leaders and operatives have undergone computer training—and hence, the group is reputed to have its own cyber unit, which is capable of hacking (at least) soft targets.

For these reasons—presence within India, detailed knowledge of the geography of at least some parts of India, strong cyber technical expertise, access to terrorist training camps in Pakistan, and financial/logistical/operational support from Pakistan's intelligence agency—the Indian Mujahideen presents a near and constant threat to India and perhaps beyond as exemplified by the 2002 bombing at the American Center in Kolkata by ARCF, a predecessor of the Indian Mujahideen.

References

Ahmed I (2005) Between moderation and radicalization: transnational interactions of Jamaat-e-Islami of India. Global Networks 5

Ahmed I (2009) Islamism and democracy in India: the transformation of Jamaat-e-Islami. Princeton University Press, Princeton

Alexander PJ (2002) Policing India in the new millennium. Allied Publisher, Delhi

Associated Press (1992) Riots sweep India after mosque razed. Associated Press December 7

Banerjee A (2004) The threat in the north east. Rediff.com April 27. http://www.rediff.com/news/2004/apr/27ariban.htm. Accessed 7 May 2013

BBC News (1998) Violence mars India's election run-up. BBC News February 15. http://news.bbc.co.uk/2/hi/56682.stm. Accessed 7 May 2013

BBC News (2012) Indian Mujahideen group banned in UK. BBC News July 4. http://www.bbc.co.uk/news/uk-politics-18717807

Biswas T (2012) Abu Jundal's passport shows him as a Pakistani national. NDTV July 3. http://www.ndtv.com/article/india/abu-jundal-s-passport-shows-him-as-a-pakistani-national-239118

Chakravarty S (2001) 'I am very bitter about being an Indian'. India Today April 2. http://indiatoday.intoday.in/story/our-madarsas-are-not-nurseries-of-terror-simi-leader-safdar-nagori/1/232611.html. Accessed 7 May 2013

Chakravarty S (2002) The dons of terror: Aftab Ansari and Omar Sheikh. India Today February 25. http://indiatoday.intoday.in/story/aftab-ansari-omar-sheikh-dons-of-terror/1/164211.html

CNN/IBN News (2007) Indian Mujahideen claims responsibility for UP blasts. CNN/IBN News November 23. http://ibnlive.in.com/news/indian-mujahideen-claims-responsibility-for-up-blasts/52882-3.html. Accessed 7 May 2013

CNN (2013) India Arrests Yasin Bhatkal, Indian Mujahideen Terrorism Suspect, August 29 2013.http://www.cnn.com/2013/08/29/world/asia/india-terrorist-arrest/index.html. Accessed August 30 2013

Cohen S (2004) The idea of Pakistan. The Brookings Institution, Washington DC

Coll S (2004) Ghost wars: The secret history of the CIA, Afghanistan, and Bin Laden, from the Soviet invasion to September 10, 2001.Penguin Press, New York

Deccan Herald (2013) Al-Umma men under scanner in TN. Deccan Herald April 24. http://www.deccanherald.com/content/328364/al-umma-men-scanner-tn.html. Accessed 7 May 2013

DNA India (2013) Rs 1 Lakh 'eidi' Gave Away Yasin Bhatkal's Nepal Hideout, 30 August 2013. http://www.dnaindia.com/india/1882139/report-rs1-lakh-eidi-to-wife-gave-away-yasin-bhatkal-s-nepal-hideout Accessed Aug 31 2013

Engineer A (1991) The bloody trail: Ramjanmabhoomi and communal violence in UP. Economic and Political Weekly January 26

Fair CC (2010) Students Islamic Movement of India and the Indian Mujahideen: an assessment Asia Policy 9

Gaikwad R (2009) Indian Mujahideen wanted bases in Maharashtra. The Hindu February 18. http://blogs.thehindu.com/delhi/?p=14383. Accessed 7 May 2013

Gargan E (1993) India bombings: gangs involved, but who else? The New York Times May 16

Government Report (2001) A home ministry report on SIMI activities. The Newspaper Today September 27. http://www.hvk.org/2001/0901/180.html. Accessed 7 May 2013

Gupta S (2011) The Indian Mujahideen: tracking the enemy within. Hachette, Gurgaon

Habibullah W (2008) My Kashmir: conflict and the prospects of enduring peace. United States Institute of Peace, Washington DC.

Hafeez K, Hafeez M (2009) Peerbhoy led media wing. Times of India/TNN February 18. http://articles.timesofindia.indiatimes.com/2009-02-18/mumbai/28008172_1_media-houses-minutes-mansoor-peerbhoy-terror-mails. Accessed 7 May 2013

Haque MM (2002) Indian Muslims forced to extremism. The Milli Gazette. http://www.milligazette.com/Archives/01112002/0111200282.htm. Accessed 7 May 2013

Haqqani H (2005) Pakistan: between mosque and military. Carnegie Endowment for International Peace, Washington DC

Hazarkia S (1992) Hindus are forced from mosque site. The New York Times December 9.

The Hindu (2012) 13/7 accused Haroon Naik met Osama, Lakhvi in Pak. ATS, The Hindu February 7. http://www.thehindu.com/news/national/137-accused-haroon-naik-met-osama-lakhvi-in-pak-ats/article2869467.ece. Accessed June 19, 2013

Hindustan Times (2008) 'SIMI, IM new faces of Lashkar-e-Taiba.' Indo-Asian News Agency/ Hindustan Times September 20. http://www.hindustantimes.com/India-news/NewDelhi/SIMI-IM-new-faces-of-Lashkar-e-Taiba/Article1-339183.aspx. Accessed 7 May 2013

Hindustan Times (2013) Accused Shahzad held guilty in Batla House encounter case, court says raid genuine. Hindustan Times July 25. http://www.hindustantimes.com/India-news/NewDelhi/Accused-Shahzad-held-guilty-in-Batla-House-encounter-case-court-says-raid-genuine/Article1-1098112.aspx?hts0021

Imtiaz S (2012) 10 years on, the Daniel Pearl kidnapping and murder case. The Express Tribune October 15. http://tribune.com.pk/story/451550/10-years-on-the-daniel-pearl-kidnapping-and-murder-case/

Indian Express (2008) Is Indian Mujahideen actually the Asif Reza commando force?, Sep 30, 2008, http://www.indianexpress.com/news/is-indian-mujahideen-actually-the-asif-reza-commando-force-/367590/

Jaleel M (2013) Two terror attacks, four sets of accused, two names in common. The Indian Express July 5. http://www.indianexpress.com/news/two-terror-attacks-four-sets-of-accused-two-names-in-common/1115309/0

Jain B (2013) A single attack that earned Jaish-e-Mohammad global attention. The Times of India February 10. http://articles.timesofindia.indiatimes.com/2013-02-10/india/37019944_1_hijack-jem-al-faran

John W (2012) The caliphate's soldiers: the Lashkar-e-Tayyeba's long war. Amaryllis Publications

Johnson TA (2009) Nasir linked to series of blasts, terror outfits. Indian Express December 21. http://www.indianexpress.com/news/nasir-linked-to-series-of-blasts-terror-outfits/556947/0. Accessed 7 May 2013

Katakam A (2003) The New Pawns. Frontline 20. http://www.frontline.in/static/html/fl2019/stories/20030926003802100.htm. Accessed 7 May 2013

King G (2004) The Most Dangerous Man in the World: Dawood Ibrahim. Chamberlain Brothers

Lal R (2004) Islam in India. In A Rabasa et al. (ED) The Muslim world after 9/11. RAND Corporation, Arlington VA

Lerman E (1981) Mawdudi's concept of Islam Middle Eastern Studies 17: 492–509

Mid-Day (2003) An interview with former SIMI president: Sayeed Khan. Mid-Day September 21. http://www.mid-day.com/news/2003/sep/64447.htm. Accessed 7 May 2013

Nanavati G and Mehta A (2008) Report by Commission of Inquiry. http://home.gujarat.gov.in/homedepartment/downloads/godharaincident.pdf

Nanjappa V (2012) How the underworld funds Indian Mujahideen's deadly plans. Rediff.com, May 26. http://www.rediff.com/news/report/how-the-underworld-funds-indian-mujahideens-deadly-plans/20120526.htm

Nasr V (1994) The vanguard of the Islamic revolution: the Jama'at Islami of Pakistan. University of California Press, Berkeley

Nasr V (1996) Mawdudi and the making of Islamic Revivalism. Oxford University Press, Oxford

NDTV News (2010a) Terror plot foiled in Mumbai, men with Pak links held. NDTV News March 14. http://www.ndtv.com/article/india/terror-plot-foiled-in-mumbai-men-with-pak-links-held-17674. Accessed 7 May 2013

NDTV News (2013) Yasin Bhatkal, alleged chief of Indian Mujahideen, arrested. NDTV News August 30. http://www.ndtv.com/article/india/yasin-bhatkal-alleged-chief-of-indian-mujahideen-arrested-411979 Accessed Aug 30 2013.

Outlook India (2008) For the record: 'The Rise of Jihad, Revenge of Gujarat.' Outlook India July 28. http://www.outlookindia.com/article.aspx?238039

Peters G (2003) Al Qaeda-Pakistani ties deepen. Christian Science Monitor March 6. http://www.csmonitor.com/2003/0306/p01s04-wosc.html

PTI (2012) Centre extends ban on SIMI for two years. Press Trust of India February 9. http://zeenews.india.com/news/nation/centre-extends-ban-on-simi-for-two-years_757453.html

Raman B (2010) The Jihadi terrorism in India: the Saudi connection. International Terrorism Monitor 624 March 1. http://ramanstrategicanalysis.blogspot.com/2010/03/jihadi-terrorism-in-india-saudi.html

Rajamohan PG (2005) Tamil Nadu: the rise of Islamist fundamentalism. Faultlines 16 January. http://www.satp.org/satporgtp/publication/faultlines/volume16/Article5.htm. Accessed 7 May 2013

Rana MA (2006) A to Z of Jehadi organizations in Pakistan. Mashal Books, Lahore

Rediff News (2011) IM front JAM under the scanner for Mumbai blasts. Rediff News July 18. http://www.rediff.com/news/report/im-front-jam-under-the-scanner-for-mumbai-blasts/20110718.htm. Accessed 7 May 2013

Rediff.com (2008b) How the Indian Mujahideen was formed. Rediff.com July 29. http://www.rediff.com/news/2008/jul/29ahd9.htm. Accessed 7 May 2013

Rediff.com (2002) Ansari admits to taking Lashkar's help. Rediff.com February 15. http://www.rediff.com/news/2002/feb/14ansari.htm. Accessed 7 May 2013

Rediff.com (1997) Let them not demolish the mosques in Kashi and Mathura or the country will face another Partition. Rediff.com December 1. http://www.rediff.com/news/dec/01al.htm.

Reidel B (2013) Avoiding Armageddon: America, India, and Pakistan to the brink and back. Brookings, Washington DC

Roul A (2010) After Pune, details emerge on the Karachi project and its threat to India. CTC Sentinel April 2010. http://www.ctc.usma.edu/posts/after-pune-details-emerge-on-the-karachi-project-and-its-threat-to-india. Accessed 7 May 2013

SATP (2013) Al Badr, South Asia Terrorism Portal. http://www.satp.org/satporgtp/countries/india/states/jandk/terrorist_outfits/AL_BADR.HTM

Sareen S (2005) The Jihad factory: Pakistan's Islamic revolution in the making. Observer Research Foundation, New Delhi

Selznick P (1952) The organizational weapon: a study of Bolshevik strategy and tactics. The RAND Corporation. http://www.rand.org/content/dam/rand/pubs/reports/2006/R201.pdf

Shahzad SS (2003) Ceasefire will not hold, with same game, new rules. South Asia Tribune November 30 - December 6. http://antisystemic.org/satribune/www.satribune.com/archives/nov30_dec6_03/opinion_saleem.htm. Accessed 7 May 2013

Sharma R (2012) SIMI backbone of terror outfits: Abu Jundal. Asian Age July 26. http://archive.asianage.com/india/simi-backbone-terror-outfits-abu-jundal-865

Sharma V (2008) DJS chief still getting Gulf aid. New Indian Express Dec. 5. http://newindianexpress.com/cities/hyderabad/article12003.ece. Accessed 7 May 2013

Sharma V (2013a) SIMI man held, brought to city. New Indian Express March 5. http://newindianexpress.com/states/andhra_pradesh/article1488744.ece. Accessed 7 May 2013

Sharma V (2013b) Maqbool says recruits streaming into Indian Mujahideen. New Indian Express March 02. http://newindianexpress.com/states/andhra_pradesh/article1484741.ece. Accessed 7 May 2013

Shashikumar VK (2008) Indian Mujahideen cover for international terror groups. CNN-IBN July 28, 2008, http://ibnlive.in.com/news/indian-mujahideen-a-cover-for-international-groups/69793-3-1.html. Accessed 7 May 2013

Siddiqi MA (2003) The SIMI I founded was completely different. Rediff.com September 2. http://www.rediff.com/news/2003/sep/02inter.htm

Sikand Y (2002) The emergence and development of the Jama at-i-Islami of Jammu and Kashmir (1940s 1990). Modern Asian Studies 36: 705–751

Sikand Y (2006) The SIMI story. Countercurrents July 15. http://www.countercurrents.org/comm-sikand150706.htm. Accessed 7 May 2013

Sikand Y (2001) Countering fundamentalism: beyond the ban on SIMI. Economic and Political Weekly October 6

Singh PK (2003) SIMI's Nagori meets Maulana too. The Pioneer July 21. http://www.hvk.org/2003/0703/166.html. Accessed 7 May 2013

Singh V (2012) Ansari trained 90 at Lashkar camp. Indian Express, June 27, 2012, http://www.indianexpress.com/news/-ansari-trained-90-at-lashkar-camp-/967114/. Accessed 7 May 2013

Spaeth A (1993) Bombay perplexed over aim of bombings. Christian Science Monitor March 15.

Subramanian TS (2000) Indicting the police. Frontline June 10–23. http://www.frontline.in/navigation/?type=static&page=archiveSearch&aid=17120490&ais=12&avol=17. Accessed 7 May 2013

Subramanian TS (1998) A time of troubles. Frontline March 7–20. http://www.frontline.in/navigation/?type=static&page=flonnet&rdurl=fl1505/15050170.htm. Accessed 7 May 2013

Subrahmanian VS, Mannes A, Shakarian J, Sliva A, and Dickerson J (2012) Computational analysis of terrorist groups: Lashkar-e-Taiba. Springer, New York

Swami P (2000) The 'liberation of Hyderabad. Frontline May 13–26. http://www.frontline.in/static/html/fl1710/17100390.htm

Swami P (2007) India, Pakistan and the secret jihad: the covert war in Kashmir, 1947–2004. Oxon, Routledge

Swami P (2008) Storm rages within SIMI. The Hindu March 11. http://www.hindu.com/2008/03/11/stories/2008031159921000.htm

Swami P (2010) Riyaz Bhatkal and the origins of the Indian Mujahidin. CTC Sentinel May 3. http://www.ctc.usma.edu/posts/riyaz-bhatkal-and-the-origins-of-the-indian-mujahidin

Swami P (2011) Indian Mujahideen manifestos attacked judiciary. The Hindu September 8. http://www.thehindu.com/news/national/article2436739.ece

Tankel S (2011) Storming the world stage: the story of Lashkar-e-Taiba. C. Hurst & Co, London

Times of India (2012a) Pakistan bomber helped 13/7 accused steal vehicles: police. Times of India June 5. http://articles.timesofindia.indiatimes.com/2012-06-05/mumbai/32055149_1_naqi-ahmed-scooters-yasin-bhatkal. Accessed 7 May 2013

Times of India (2012b) ISI created Indian Mujahideen to spread terror in India, says anti-terrorism squad. Times of India June 2. http://articles.timesofindia.indiatimes.com/2012-06-02/india/31982947_1_bomb-blasts-chargesheet-anti-terrorism-squad. Accessed 7 May 2013

Times of India (2012c) Dawood gang member acquitted in kidnapping case. Times of India March 31. http://articles.timesofindia.indiatimes.com/2012-03-31/kanpur/31266370_1_dawood-gang-member-kidnapping-fazlur-rehman. Accessed June 19 2013

Tiwary D (2012) Indian Mujahideen founder Amir Reza Khan still hiding in Pak: NIA. Times of India November 24. http://articles.timesofindia.indiatimes.com/2012-11-24/india/35333661_1_im-men-luxembourg-nia-officer. Accessed 7 May 2013

U.S. Department of State (2011) Terrorist designations of the Indian Mujahideen. U.S. Department of State September 15. http://www.state.gov/r/pa/prs/ps/2011/09/172442.htm

Verhovek S (1992) Frail social compact is broken in Bombay. New York Times December 10

Zee News (2009) All three LeT terrorists convicted for 2003 Mumbai blasts. Zee News July 28. http://zeenews.india.com/news/nation/all-three-let-terrorists-convicted-for-2003-mumbai-blasts_550599.html. Accessed 7 May 2013

Chapter 3
Temporal Probabilistic Behavior Rules

Abstract This chapter describes the syntax and semantics of Temporal Probabilistic (TP) behavioral rules used throughout the book to describe the behavior of the Indian Mujahideen. The chapter describes the intuition behind TP-rules and their formal syntax and meaning, and describes an algorithm used to derive the TP-rules automatically from data about IM.

This chapter focuses on some of the technology underlying this book. Readers whose sole interest is in behavioral models of IM and in policies to rein in violence carried out by IM may skip this chapter without any loss of relevant material.

The first mathematical models of behaviors of terror groups were expressed via Stochastic Opponent Modeling Agent rules (SOMA-rules) (Khuller et al. 2007; Simari et al. 2012). A SOMA-rule is an expression of the form

> When condition C is true in the environment in which terror group G operates, there is a probability of p % that terror group G will take action A at intensity level I.

SOMA-rules have been used extensively to generate automatically expressed rules about the behaviors of terror groups such as Hezbollah (Mannes et al. 2008a, b; Mannes and Subrahmanian 2009), Hamas (Mannes et al. 2008a, b) and LeT (Mannes et al. 2011) and demonstrate considerable expressive and predictive power. As an excellent example, Mannes et al. (2008a, b) predicted in April 2008 Hezbollah's behavior in the first part of 2009 prior to scheduled Lebanese elections. In November 2008, a reporter for the Beirut *Daily Star* made skeptical comments regarding the predictions and included comments from Hezbollah about the predictions. A Lebanese academic political scientist had similar comments. Hezbollah then behaved exactly as predicted in the first half of 2009 despite knowing about the predictions of Mannes et al. (2008a, b).

A slight variant of this chapter appeared as Chap. 3 of Subrahmanian et al. (2012).

V. S. Subrahmanian et al., *Indian Mujahideen*, Terrorism, Security, and Computation, DOI: 10.1007/978-3-319-02818-7_3,

SOMA-rules however can occasionally be a bit strange. An example of such a SOMA-rule describing the behavior of IM that we derived automatically is given below.

> IM carries out bombings with an 87.5% probability during a given month if:
>
> - IM issued a claim of responsibility during that month
>
> *Support* = 7
> *Inverse Probability* = 1

This SOMA-rule is easily read and understood. The reader however may raise questions about cause and effect. As both events, "IM carried out bombings" and "IM issued a claim of responsibility", occurred during the same month, we are unable to tell whether the bombings preceded the claims of responsibility or vice versa. If the claim of responsibility was cleanly asserted prior to that month's bombing (and was therefore associated with a bombing from a prior month), then claims of responsibility might signal future bombings. On the other hand, if the bombing during the current month preceded the claim of responsibility, then the claim of responsibility might have been for the bombing carried out that month—this yields no predictive power. The rule would be significant and interesting in the first case but not relevant predictively in the second. The problem with SOMA-rules is that it is often *hard to tell* which of the two cases is applicable. When the events described in the "if" part of the SOMA-rule clearly do not overlap with the events occurring in the "then" part of the same rule, one can confidently put forth a significant correlative relationship between the "if" and the "then" parts of the rule. But when there is overlap in time, as in the case described above, one cannot tell.

A further problem with SOMA-rules is that they do not include a temporal element. In the real world, terror groups do not always react instantaneously to events in their environment. They deliberate internally, they develop operational plans, they plan logistics to support their plans, and they finally execute or abort them. All of this takes time, leading to temporal delays that SOMA-rules do not model.

This chapter, describes *temporal-probabilistic rules* (TP-rules) introduced by Dekhtyar et al. (1999) that avoid both these problems. Clearly articulating *delays* between the time when a terror group's environment satisfied a given condition and a *subsequent* time when the terror group took a given action[1] excludes the

[1] Our range was a 1–5 month delay, though longer delays are possible and should be studied.

possibility that the "if" part of a rule and the "then" part of the rule could have occurred simultaneously.

This chapter starts with a description of our data structure for IM, followed by a description of the syntax of TP-rules and then a description of the algorithm used to derive TP-rules.

3.1 Database Schema

The data about IM are represented as a relational database table; the rows correspond to *months* and the columns correspond to two kinds of *attributes*. The identical data format is used in our CMOT project (Shakarian et al. 2012) to collect information about a host of terror groups, not just IM. Other terror groups for which data have been collected using the same representation include LeT (Subrahmanian et al. 2012), JeM, SIMI, and the Forces Democratiques de Liberation du Rwanda (FDLR). The FDLR's leaders are widely held responsible for the 1994 genocide of Tutsi civilians in Rwanda.

Environmental *attributes* refer to aspects of the environment within which IM functioned. These include attributes relating to:

- The internal structure of IM;
- Information on the internal activities of IM (e.g., inter-organizational conflict);
- Information about IM's resources including financial, military, and political support, as well as similar kinds of support by the diaspora and foreign states;
- Information about IM's ongoing campaigns (e.g., news/media campaigns);
- Information about IM's facilities (e.g., training camps);
- Information about IM gatherings (e.g. conferences);
- Information about IM's grievances (e.g. allegations of poor treatment of Muslims in India); and
- Information about actions, both supportive and adversarial, taken by external actors such as the U.S., India, and Pakistan towards IM; they include bans of IM, kills/arrests of IM personnel, issuance of arrest warrants, and/or freezing assets.

The second class of attributes in modeling the behavior of IM are *action attributes* describing the intensity of various types of actions taken by IM. Actions are typically described by location (national or transnational), type of action (e.g., fedayeen attack versus hijacking versus armed clash), and type of target (e.g., security installation versus transportation facility). The research conducted involved the following action attributes:

- Armed attacks
- Bombings
- Attacks on public sites
- Suicide attacks
- Hijackings

- Abductions/kidnappings
- Attacks on security targets (e.g., police stations or military bases)
- Attacks on transportation targets (e.g., airports, train stations)
- Attacks against civilians on the basis of ethnicity or religion (e.g., against Hindus, Christians, or Jews)
- Attacks on government facilities (including government offices) not related to security forces
- Attacks on transportation targets (e.g., the 2006 Mumbai train attacks)
- Armed clashes with different types of security forces
- Simultaneous attacks on multiple, geographically disparate sites.

As in standard relational databases, each attribute (environmental or action) has a precisely stated *domain*. In most work related to the modeling of ethnic groups or terror groups, the domains are binary, meaning the group either performed a given action or not.

In contrast to most work, our IM database includes frequency and personnel counts, e.g., the number of armed clashes during a given month between IM and security forces, the number of IM leaders arrested during a given month, In other words, the database captures variations in intensity—other projects modeling terror groups were unable to do so.

Appendix A provides a more detailed description of our data and methodology.

3.2 TP-Rule Syntax

TP-rules use a form of logic programming and computational logic to express rules that have both a temporal and a probabilistic aspect. TP-rules are a variant of Generalized Annotated Programs (Kifer and Subrahmanian 1992), first introduced by Dekhtyar et al. (1999) and later studied and extended to *annotated probabilistic temporal* (APT) logic programs by Shakarian et al. (2011).

Every attribute p in the data set described in Sect. 3.1 corresponds to a *unary predicate symbol*. The argument attached to a unary predicate symbol can either be a value v from the domain of the attribute p or a variable X ranging over this domain. v and X are *terms* over the domain of p.

If t is a term over the domain of predicate symbol p, then $p(t)$ is called an *atom*. When t is in the domain of p, $p(t)$ is called a *ground atom*. When p is an action (resp. environmental attribute), $p(t)$ is an *action atom* (resp. *environmental atom*).

For example, suppose *suicide-attack* is an action attribute whose domain is the non-negative integers ($\geq$0). Then *suicide-attack* (3) is a ground atom referring to three suicide attacks. Similarly, *suicide-attack* (10) is a ground atom referring to 10 suicide attacks, while *suicide-attack* (X) is a ground atom that can be instantiated to a ground atom where the variable X takes any a concrete value—representing any desired number of suicide attacks.

Suppose $t' \geq 0$ is either a time point (non-negative integer) or a variable ranging over non-negative integers, and $p' \varepsilon [0, 1]$ is a probability. Then $[t', p']$ is called a *temporal probabilistic annotation* (*tp-annotation*).

For example, [10, 0.7] refers to a probability of 70 % or more of some (unspecified event) occurring 10 time units after a fixed time.

If $p(t)$ is an atom and $[t', p']$ is a tp-annotation, then $p(t)$:$[t', p']$ is a *temporal probabilistic annotated atom* (*tp-annotated atom*, for short).

For instance, *suicide-attack* (3):[2, 0.8] is a tp-annotated atom that says that three suicide attacks will occur with 80 % probability two time units after a fixed time.

If $p(t)$ is an (action, environmental) atom and $[t', p']$ is a tp-annotation, then $p(t)$:$[t', p']$ is a (resp. action, environmental) *tp-annotated atom.* $p(t)$:$[t', p']$ is *ground* if and only if $p(t)$ is ground—otherwise it is *non-ground.*

If X is a variable over non-negative integers and Y is a term over non-negative integers, then $X = Y$, $X \leq Y$, $X < Y$, $X \geq Y$ are all comparison atoms.

If $A_1, \ldots, A_n$ are environmental atoms or comparison atoms and $p(t)$:$[t', p']$ is a tp-annotated action atom, then

$$p(t) : [t', p'] \leftarrow A_1 \& \ldots \& A_n$$

is a *temporal-probabilistic rule* (TP-rule) that intuitively states that if $A_1, \ldots, A_n$ are all true within IM's environment at a given time τ, then $p(t)$ is true at time $(\tau + t')$ with probability p'. $p(t)$ is called the *head* of this TP-rule, while $A_1 \& \ldots \&$ A_n is called the *body*.

On the other hand, consider the TP-rule

TP-Rule PS-3. IM attacks civilians two months after months in which:

- IM made claims of responsibility for certain terrorist attacks.

  Support = 4
  Probability = 100%, *Inverse Probability* = 100%,
  Negative Probability = 0%

which we discuss in further detail in Chap. 4. It can be written in TP-rule syntax as:

$$civilian - attack\,(1) : [2, 1] \leftarrow claim\_of\_responsibility\,(1).$$

This rule says that IM carries out one civilian attack with 100 % probability 2 months after a given period of time in which IM issued a claim of responsibility for an attack.

Note that there are also data in our IM dataset that looks not at the number of events but qualitatively scales the intensity of events.

Another example of a TP-rule that we derived about bombings by IM is given in Chap. 5 and says:

TP-Rule BOMB-1. IM carries out bombings two months after months in which:

- Between 1 and 21 IM members were arrested.

Support = 5
Probability = 100%, *Inverse Probability* = 100%,
Negative Probability = 0%

This TP-rule can be formally expressed as:

$$bombing(1) : [2, 1] \leftarrow govt - arrest(\mathrm{X})\,\&1 \leq \mathrm{X} \leq 21.$$

This rule uses quantitative information in the TP-rule body and quantitative information in the rule head. It says that when the government arrests X IM personnel and X is between 1 and 21, then IM carries out 1 or more bombings 2 months later with 100 % probability.

In this book, we do not go into the details of the logical and probabilistic semantics of TP-rules; this book describes modeling the IM using TP-rules. Nor do we go into the details of the algorithms used to reason with the TP-rules. Such details are presented in detail by Dekhtyar et al. (1999) and Shakarian (2012).

3.3 SOMA-Rules

Subrahmanian and Ernst (2009) have developed methods to automatically extract "SOMA-rules" *without time* from data of the kind described in Sect. 3.1. SOMA-rules are probabilistic rules that use the syntax of probabilistic logic programs (Ng and Subrahmanian 1993).

As SOMA-rules are not used extensively in this book, we review them very briefly here. An annotated atom is an expression of the form A: $[v_1,v_2]$ where v_1,v_2 are reals in the [0,1] interval. If $A_1, \ldots, A_n$ are environmental atoms or comparison atoms and $p(t)$:$[v_1,v_2]$ is an annotated atom, then

$$p(t) : [v_1, v_2] \leftarrow A_1\&\ldots\&A_n$$

is a SOMA-rule. Intuitively, this rule says that if A_1 &... & A_n are all, then $p(t)$ is true with probability in the interval $[v_1,v_2]$. $p(t)$ is called the *head* of this SOMA-rule, while A_1 &... & A_n is called the *body*.

SOMA-rules have been used extensively to reason about the behavior of terror groups such as Hezbollah, Hamas, and LeT. In addition, SOMA-rules have been extracted automatically from data from about 50 groups. In a completely different setting, SOMA-rules have been used to extract relationships between socio-political-economic data and educational outcomes in 221 countries as well as about the relationship between economic variables and systemic banking failures (Minoiu et al. 2013).

3.4 Extracting SOMA-Rules Automatically

In this section, we briefly describe the method of Subrahmanian and Ernst (2009) to extract SOMA-rules automatically. This method has been extended to automatically extract TP-rules. The extraction of TP-rules is described in Sect. 3.5.

The Subrahmanian–Ernst (SE for short) algorithm generates rules whose bodies consist of *bi-conjuncts* which are expressions of the form $p(X)$ & $L \leq X \leq U$. A *bi*-SOMA-rule is an expression of the form

$$p(t) : [v_1, v_2] \leftarrow B_1\&\ldots\&B_n$$

where each B_i is a *bi*-conjunct. We call B_1 &... & B_n a *bi*-body.

Each *bi*-SOMA-rule is also an ordinary SOMA-rule in so far as the SE algorithm generates *bi*-SOMA rules automatically.

Given two *bi*-bodies C_1 and C_2 of environmental atoms, we say that C_1 and C_2 are *equivalent* (with respect to a given dataset such as our IM dataset) if two conditions hold:

1. The set of months in which C_1 is true for IM is identical to the set of months in which C_2 is true, i.e. the two conditions always co-occurred, and
2. C_1 and C_2 involve exactly the same environmental atoms.

The equivalence relation above induces equivalence classes on the set of all *bi*-bodies. From each equivalence class, a *canonical member* is selected to represent that equivalence class. As all *bi*-bodies within an equivalence class are equivalent in terms of the environment attributes referenced and in terms of their truth value, any member of the class is representative of the class as a whole.

Canonical members are required to be *tight*. We do not define *tight* formally here, but instead give an example. Suppose an equivalence class consists of the *bi*-bodies C_1, C_2, ..., C_n. Suppose C_i contains the *bi*-conjunct $p(X)$ & $L_i \leq X$ - $\leq U_i$. Clearly all the C_i's must contain such *bi*-conjuncts; otherwise they would not be in the same equivalence class. The canonical representative of the class must therefore contain the *bi*-conjunct

$$p(X)\ \&\ \min\{L_i | i = 1,\ldots,n\} \leq X \leq \max\{U_i | i = 1,\ldots,n\}.$$

The same principle applies to all other attributes occurring in the equivalence class $\{C_1, C_2, \ldots, C_n\}$.

The dimension of a bi-body is the number of attributes in it. The SE algorithm defines a "simpler-than" ordering on bi-bodies as follows.

C_1 is "simpler than" C_2, denoted $C_1 \gg C_2$iff C_1 has the same or fewer number of attributes in it than C_2.

Intuitively, the "simpler than" relationship looks at the number of attributes in *bi*-bodies. This however is not enough.

Given a specific action variable value *p*(*t*) that we want to predict, the SE algorithm then defines a more general ordering on conditions. $C_1 \gg C_2$ iff:

$$C_1 \gg C_2 \text{ and}$$

$$\text{Conf}(C_1) \geq \text{Conf}(C_2) \text{ and}$$

$$\text{Sup}(C_1) \geq \text{Sup}(C_2)$$

where Conf(*C*) is the confidence of condition *C* w.r.t. predicting *p*(*t*), i.e.,

$$Conf(C) = \frac{\textit{number of months when C was true and } p(t) \textit{ was true}}{\textit{number of months where C was true}}$$

and Sup(*C*) is the support of *C*, which is the numerator of the above formula.

Intuitively, for a *bi*-body C_1 to be better than a bi-body C_2 (i.e., for $C_1 \gg C_2$ to hold), we require that C_1 not only be simpler than C_2 but also that C_1 have higher support and confidence than C_2.

Throughout this chapter, we assume that we only consider *bi*-bodies of dimension *p* or less for some fixed *p*. Intuitively, *p* denotes how large a *bi*-body can be and can be set by the system. In our rule extractor, we set *p* to 3 which means that the body of any SOMA or TP-rule we consider will have at most 3 *bi*-conjuncts in it.

Given a *bi*-body *C*, the *up-set* of *C*, denoted *up*(*C*), is the set of all *bi*-bodies of dimension *p* or less such that C' is simpler than *C*. Formally, $up(C) = \{C' \mid C'$ is a tight *bi*-body and C' is of dimension *p* or less and $C' \gg C\}$.

We can now define a way of iteratively computing such *bi*-bodies.

$$\text{Tp} \uparrow 1 = \{C|C \text{ is tight and dimension}(C) \leq p \text{ and up}(C) = \{\}\}.$$

$$\text{Tp} \uparrow (\text{i}+1)$$
$$= \{C|C \text{ is tight and dimension}(C) \leq p \text{ and up}(C) \text{ is a subset (or equal to) Tp} \uparrow \text{i}\}.$$

When computing Tp↑k, the SE algorithm uses something called a *condition graph* (COG) whose vertices are (labeled with) tight *bi*-bodies.

There is an edge from vertex *u* to vertex *v* if :

1. *u*·BiBody $\gg$ *v*·BiBody and
2. There is no vertex *w* such that *u·BiBody* $\gg$ *w·BiBody* $\gg$ *v·BiBody*.

Each vertex in a COG has a level. A vertex with in-degree 0 has level 0; otherwise, the level of a vertex v is $1 + \max\{\text{level}(u) \mid \text{there is an edge from } u \text{ to } v \text{ in the COG}\}$.

Rather than building a COG completely, the SE algorithm uses COGs to generate Tp↑k efficiently when a specific outcome (e.g., *civilian-attack*(1)) is specified as input, i.e., we want to find rules that effectively predict when IM launches attacks against civilians with intensity of 1.

The following algorithm contains several procedures that are briefly summarized below.

The **Build-Data-Struc**() procedure builds a data structure that contains all rows in the IM database that satisfy the desired Outcome condition (in the case of the SatisfyOutcome line) or do not satisfy the desired Outcome condition (in the case of the notSatisfyOutcome line).

The **GenerateTightBi-bodies**() function generates all tight *bi*-bodies associated with attributes in φ using the table DB containing the IM data. For each such tight bi-body v generated, it creates a record with the fields *bibody* (specifying the *bi*-body), *conf* (specifying the confidence of the rule $Outcome \leftarrow v$) and *sup* (specifying the support).

The **InsertCOG** function inserts the vertex v into the COG if the level of the vertex would be K or less. It then updates the neighbors of the vertex so that their levels are appropriately reset.

The **Compute Tp ↑ k** algorithm finds all *bi*-bodies in Tp ↑ k and these *bi*-bodies form the bodies of rules whose head is *Outcome*. By invoking this function for all desired action attributes (and associated values) associated with IM, we were able to derive all possible rules that satisfied desired support and confidence levels.

Note that once the computation of the **Compute Tp ↑ k** algorithm is completed—but immediately before the **return** statement at the end of the code, it is possible to subject each of the derived bi-bodies in **ExtractBiBody** (COG) to any additional statistical test that a user wishes to subject it to. This allows our algorithm to be used in conjunction with third party desired statistical tests with little modification.

```
Compute Tp↑k(DB,ENV,Outcome,p,k)
COG = NIL; (* initially there are no vertices in the COG *)
Foreach combination φ of p or less environmental attributes do
SatisfyOutcome = Build-Data-Struc(DB,ENV, φ,Outcome);
NotSatOutcome = Build-Data-Struc(DB,ENV, φ,~Outcome);
   (* SatisfyOutcome contains the projection of DB on attributes in φ
for months that satisfy the desired outcome; NotSatOutcome does the
same for months that do not *)
TightBi-Bodies= GenerateTightBi-Bodies( φ,SatisfyOutcome);
(* generates tight bi-bodies associated with attributes in φ and
 computes their support *)
Foreach v in TightBi-Bodies do
     (* v is record with fields bibody,conf,sup *)
     numNotOutcome=
              CountQuery(v.bibody,NotSatisfyOutcome)
     (* finds number of months in which v.bibody was true
      but Outcome did not occur *)
     v.confidence = v.support/(v.support + numNotOutcome);
     COG = InsertCOG(v,COG,K);
endfor;
endfor;
return ExtractBiBody(COG)
end
```

3.5 Automatically Extracting TP-Rules

In order to automatically extract TP-rules associated with a time offset of j months, we applied the Compute Tp ↑ k algorithm to a new database DB' (instead of the actual IM database DB).

DB' can be calculated from DB as follows:

- Set $DB1$ to DB.
- Eliminate all rows in $DB1$ associated with the first j months.
- For each month m in $DB1$, replace all environment attribute values $m \cdot E$ by the environmental attribute value $(m - j) \cdot E$ from the original table DB.
- DB' is the result.

What this manipulation does is "pretend" that the environment in month m is actually the environment in month $(m - j)$ and thus, it "fools" the **Compute Tp ↑ k** algorithm into computing the correct time-offset rules.

We also note that if an application or user wants TP-rules to derive additional statistical conditions (e.g. requiring that p-values be less than 0.03), then all that needs to be done is to pass the results generated by the **Compute Tp ↑ k** algorithm through an additional filter that eliminates all TP-rules returned by the **Compute Tp ↑ k** algorithm that do not satisfy this statistical condition.

3.6 Conclusion

The system computed TP-rules for time offsets 1, 2, 3, 4, and 5 months for the entire IM dataset.

Computing all TP-rules that satisfy various support and confidence criteria is challenging and expensive (from a compute time perspective). Nonetheless, we were able to generate over 37,000 TP-rules. Many of these rules were either uninteresting or repetitive—we manually went through all TP-rules to find the interesting ones. Chapters 4–7 describe the most interesting TP-rules discovered pertaining to the following types of "bad acts" by IM:

- Attacks on public sites;
- Bombings;
- Simultaneous/timed attacks;
- Total number of people killed.

In contrast to our methodology applied to LeT (Subrahmanian et al. 2012), we lowered our "support" thresholds for IM in so far as IM has been in existence for a shorter period of time than LeT and has been responsible for fewer terrorist attacks.

References

Dekhtyar A, Dekhtyar M, Subrahmanian VS (1999) Temporal pobabilistic logic programs. Proc 1999 Intl conf on logic programming, November

Khuller S, Martinez V, Nau D, Simari G, Sliva A, Subrahmanian VS (2007) Computing most probable worlds of action probabilistic logic programs: scalable estimation for 1,030,000 worlds. Annals of mathematics and artificial intelligence 51:295–331

Kifer M, Subrahmanian VS (1992) Theory of generalized annotated logic programming and its applications. J logic programming 12:335–368

Mannes A, Subrahmanian VS (2009) Calculated terror, Foreign Policy (online edition) December 15. http://www.foreignpolicy.com/articles/2009/12/15/calculated_terror?page=full

Mannes A, Michaell M, Pate A, Sliva A, Subrahmanian VS, Wilkenfeld J (2008a) Stochastic opponent modeling agents: a case study with Hezbollah. Proc 2008 first intl workshop on social computing, behavioral modeling and prediction April 1–2 Springer Verlag, Phoenix

Mannes A, Sliva A, Subrahmanian VS, Wilkenfeld J (2008b) Stochastic opponent modeling agents: a case study with Hamas. Proc 2008 intl conf on computational cultural dynamics, September

Mannes A, Sliva A, Subrahmanian VS (2011) A computational enabled analysis of Lashkar-e-Taiba attacks in Jammu and Kashmir. Proc 2011 IEEE European intelligence & security informatics conference, September

Minoiu C, Kang C, Subrahmanian VS, Berea A (2013) The role of financial connectedness in predicting crises, 2013 Conference on Interlinkages and Systemic Risk, July 2013, Ancona, Italy

Ng R, Subrahmanian VS (1993) Probabilistic logic programming, information and computation, 101:150–201

Shakarian J (2012) The CMOT codebook. LCCD, University of Maryland Institute for Advanced Computer Studies, University of Maryland, College Park, MD. Extended and revised by Schuetzle, B and Nagel, M in 2012

Shakarian P, Parker A, Simari G, Subrahmanian VS (2011) Annotated probabilistic temporal logic, ACM transactions on computational logic 12

Shakarian P, Simari GI, Subrahmanian VS (2012) Annotated probabilistic temporal logic: approximate fixpoint implementation. ACM transactions on computational logic 13

Simari G, Martinez V, Sliva A, Subrahmanian VS (2012) Focused most probable world computations in probabilistic logic programs. Annals of mathematics and artificial intelligence 64:113–143

Subrahmanian VS, Ernst J (2009) Method and system for optimal data diagnosis. U.S. patent No. 7474987 6 January 2009

Subrahmanian VS, Mannes A, Shakarian J, Sliva A, Dickerson J (2012) Computational analysis of terrorist groups: Lashkar-e-Taiba. Springer, New York

Chapter 4
Targeting Public Sites

Abstract Indian Mujahideen has carried out numerous attacks targeting public sites such as markets, sports stadiums, and hospitals. This chapter focuses on the circumstances under which IM has carried out these attacks and identifies key aspects of IM's environment correlated with such attacks.

Over the years, Indian Mujahideen has targeted numerous public sites including open-air markets and sports stadiums. In particular, IM has made repeated efforts to target markets.

In much of India, open-air markets consist of small streets or rectangular public areas with shops lining the streets or sides of the rectangle. These market streets are often densely packed with shoppers, especially on holidays such as Diwali (the "festival of lights" celebrated by Hindus) or Eid (celebrated by Muslims). With the exception of a few low ranking policemen, security at such markets is almost non-existent, a phenomenon IM has consistently exploited in order to target large numbers of innocent civilians. In fact, as shown in Fig. 4.1, IM attacks on public places claimed over 250 lives and injured over 900 innocent people between 2005 and 2010. These numbers do not include the counts of people killed in attacks on transportation networks, which will be discussed in another chapter. Though it is difficult to precisely allocate blame, IM is one of the principal suspects in the July 2006 Mumbai train attacks.[1]

---

[1] The other principal suspect in the 2006 Mumbai train attacks is LeT (Tankel 2011). IM and LeT are closely aligned; there is little question IM received extensive support from LeT and, at least indirectly, its ISI sponsors (Swami 2008). It is believed that on some attacks LeT and IM have collaborated closely (Tankel 2011). Indian security officials at times have stated that IM is little more than a front for LeT (Tripathi 2008), and there are many analysts who support this view (Roul 2010). But some analysts (e.g., Fair 2010) assert that Indian officials and media have attributed terror attacks to LeT rather than acknowledge the domestic terrorism problem. As it is difficult to know the exact details of the IM-LeT relationship, particularly as terrorist groups often purposefully operate behind fronts and proxies, in this book, IM is considered responsible for attacks when there is substantive evidence to support allegations of their involvement, even if such evidence is not 100 % incontrovertible.

V. S. Subrahmanian et al., *Indian Mujahideen*, Terrorism, Security, and Computation, DOI: 10.1007/978-3-319-02818-7_4,

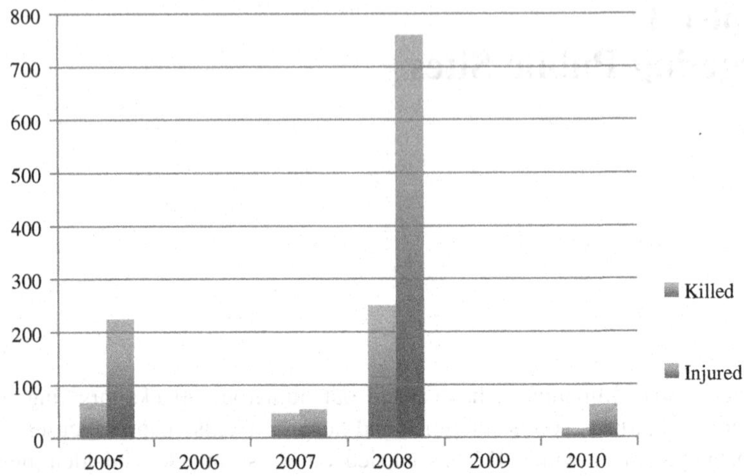

Fig. 4.1 Individuals killed or injured from IM attacks on public sites, estimated from media sources and based on attacks reasonably attributed to IM[2]

Figure 4.1 shows 2008 as the most active year for IM attacks on public sites, which is in accord with 2008 as IM's most active year overall. However, 2006 was another year in which IM was extremely active but their attacks that year targeted public transportation in Mumbai and religious sites in Varanasi, both of which were coded in different variables. IM has continued to carry out attacks on public sites in each year since 2010 (when coding was completed).

Examples of attacks on public sites widely believed to be perpetrated by IM include but are not limited to the following:

- On October 29, 2005, IM was believed to have been behind a pair of twin blasts in two markets in Delhi that led to a large number of casualties—67 dead and 225 injured (Gupta 2011). These blasts occurred in close proximity to both the major Hindu festival of Diwali and the Muslim Eid-ul-Fitr festival when large numbers of shoppers flock to markets to carry out their holiday shopping.
- On August 25, 2007, IM was largely believed to have been behind twin blasts in the central Indian city of Hyderabad, leading to 42 dead and 54 injured (Gupta 2011). Moreover, on this occasion, two other unexploded devices were discovered and neutralized by Indian police.
- IM formally announced its existence with a series of nearly simultaneous bombings at courthouses in three cities in Uttar Pradesh (Varanasi, Faizabad, and Lucknow). The bombs killed 14 people and injured over 50. IM claimed the attack in an email that was sent moments before the bombs detonated, stating

[2] There have been attacks since and other attacks not covered in the codebook, either because they were not known or were not determined to be IM operations.

that courthouses were targeted because Indian lawyers had beaten one accused terrorist and refused to defend another (Gupta 2011).

- On May 13, 2008, IM is believed to have conducted a massive coordinated attack in Jaipur, Rajasthan. Within a space of 30 min, nine bombs exploded in seven markets around Jaipur, leading to 60–80 deaths and 100–216 injuries. IM claimed responsibility for the attack by email (Ramesh 2008).
- On July 25, 2008, within a period of 90 min, a similar set of eight blasts reverberated through India's cyber capital, Bengaluru. Though the blasts were low intensity detonations and a few of the devices malfunctioned, two people were killed and seven or eight people were badly injured. Though LeT claimed responsibility, many sources believe that IM was responsible (Gupta 2011).
- On the next day, July 26, 2008, the commercial city of Ahmedabad in the economically booming western state of Gujarat was hit with 20 low-intensity blasts, including one at a trauma center where families of victims of the initial blasts had gathered. Over 30 were killed. Indian police defused another 20 bombs that day. An email sent by IM minutes before the spate of bombings claimed responsibility for the attacks (Gupta 2011), making it likely that IM was in fact responsible for these attacks.
- On September 13, 2008, in Delhi's Gaffar Market, IM detonated five explosive devices that killed 26–30 individuals while injuring between 100 and 130. An email claiming responsibility was sent by IM using a hacked computer account while the blasts were occurring (Tripathi 2008).
- On February 13, 2010, a blast rocked a bakery in the western industrial city of Pune in Maharashtra. Popular with foreigners in the area, the bakery blast claimed 17 lives and injured over 54 individuals. IM and LeT were both believed to be responsible for this attack (Roul 2010).
- On April 17, 2010, terrorists again targeted the city of Bengaluru. A bomb went off at the popular M. A. Chinnasamy Stadium where international cricket matches are frequently held. Though no one is believed to have died in this attack, eight people were injured. No group claimed responsibility, but IM is believed to be responsible (Fernandes 2012).

The LCCD's Stochastic Temporal Analysis of Terrorist Events (STATE) system derived a large number of TP-rules associated with IM's attacks on public sites and public infrastructure. These rules specifically identified the following key variables with such attacks by IM:

- *Arrests of IM Personnel.* STATE found that when IM members were arrested, IM attacks on public sites occurred 1, 2, and 3 months later. This connection is similar to rules derived about the behavior of IM's close Pakistani collaborator, LeT (Subrahmanian et al. 2012). LeT was also found to have carried out attacks of different types in India within a few months of arrests of their personnel.
- *Communications about their Campaign.* TP-rules indicated that IM is also highly likely to carry out attacks on public sites within 2 months of making a statement about their operational campaign. As an underground organization, IM does not carry out overt communications campaigns but rather includes

manifestos of its grievances, its capabilities, and its intentions for further attacks in its (usually emailed) claims of responsibility (Swami 2008). When these manifestos specifically discuss potential future attacks, IM often carries out its threats. These manifestos effectively warn when more violence is coming.

- *Communication and Claims of Responsibility.* In contrast to LeT, IM is likely to carry out attacks on public sites within 2 months of claiming responsibility for an attack. For example, IM announced its existence to the world with an e-mailed claim of responsibility after a multiple bombing in Varanasi in 2007. This finding correlates with the string of bombings targeting markets in May, July, and September of 2008, each of which was accompanied by a claim of responsibility.
- *Communications and Conferences.* Another finding of this study is that when IM holds conferences, there is a high probability that within 4 months, they will carry out attacks on public sites. As an underground organization, IM members need to communicate, but must evade detection. Secret conferences are essential for the organization to plan campaigns. At a series of secret meetings in early 2008, IM initiated and planned the series of bombings occurring later that year (TNN 2008).
- *Membership in Other Non-State Armed Groups.* The data mining algorithms found an association between situations when IM's members also belonged to other non-state armed groups (NSAGs) and their propensity to carry out attacks on public sites within 2–4 months afterwards. Shared memberships between IM and other NSAGs facilitate the sharing of organizational knowledge about how to conduct terror attacks. Likewise, shared members between different organizations may lead to an expanded logistics network including safe houses and supplies for explosive devices. This finding is similar to a finding about LeT (Subrahmanian et al. 2012)—when LeT is providing support to other Islamist groups, they are far more likely to carry out various kinds of attacks primarily directed at Indian targets.
- *Diplomatic Relations Being Entertained.* It has long been suspected that elements within the Pakistani military sabotage peace talks between the Indian government and the civilian components of the Pakistani government. Though past work did not show a connection between increasing LeT attacks and warming relations between India and Pakistan, we did find evidence of such an association with respect to IM. Specifically, significant progress in Indo-Pakistani diplomatic relations seem to be followed five months later by IM attacks which poison the atmosphere for peace talks. This raises the possibility that individuals within the Pakistani military use IM to sabotage peace talks. The role of the Pakistani military in IM operations needs further investigation.

In short, this study finds both similarities and differences between the circumstances when LeT carries out attacks on India and the circumstances when IM carries out attacks on public sites in India. Our findings lend credence to the view that IM functions as a sort of auxiliary, supporting force for LeT, carrying out attacks to support ongoing LeT operations, as well as carrying out attacks when a more than arms-length deniability is sought by elements of the Pakistani military.

4.1 Attacks Against Public Sites and Arrests of Indian Mujahideen Personnel

The study found considerable evidence for the hypothesis that arrests of IM personnel are followed within 1, 2, or 3 months later by attacks on public sites.

TP-rule (PS-1) describes such a finding in the case of 1-month time offsets.

TP-Rule PS-1. IM attacks public sites one month after months in which:

- The Indian Government arrested between 1 and 21 IM personnel.

Support = 4
Probability = 100%, *Inverse Probability* = 100%,
Negative Probability = 0%

This TP-rule has less support than similar TP-rules that were extracted by the data mining engine for LeT (Subrahmanian et al. 2012), primarily because IM has been in existence for much less time than LeT and has carried out far fewer attacks. As a consequence, the standards we used for deriving TP-rules with regard to IM were weaker than those used with respect to LeT where a minimal support of ten (as compared to four for IM) was required for inclusion in this study.

We derived an identical rule when we looked at 2-month offsets. Additionally, when we looked at 3-month offsets, an almost identical rule held, the only change being a lower support of three.

4.2 Attacks Against Public Sites and Communications About Their Campaign

When IM communications include statements about their operational campaign and discuss future attacks, there is a high probability of their launching attacks on public sites within 2 months as shown by the following TP-rule.

Figure 4.2 below highlights the relationship, with the 2-month delay, between IM statements about their terror campaign and attacks on public sites, for IM attacks since 2008.

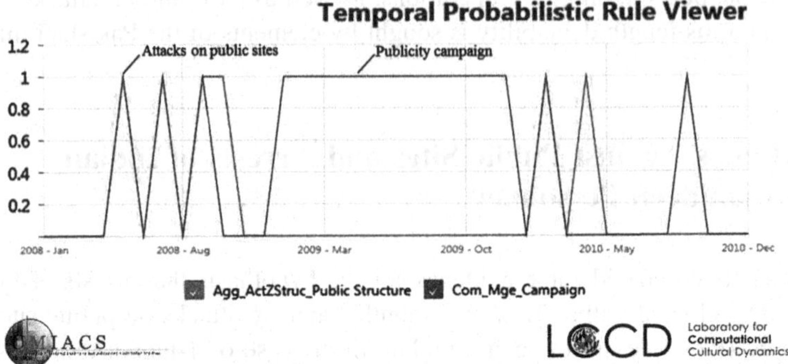

Fig. 4.2 Temporal relationship between attacks on public structures and communications campaigns carried out 2 months earlier

Unlike LeT or Hezbollah and Hamas and other terrorist groups that can operate openly, IM has fewer options for promulgating its message to its followers, potential followers, and to its enemies. Thus, the emailed claims of responsibility are in-depth manifestos describing the organization's ideology, grievances, capabilities, and strategy. Minutes before the September 2008 Delhi blasts, IM issued a 13-page email that included verses from the Koran, a record of the injustices suffered by India's Muslims, an extensive critique of the Bharatiya Janata Party (BJP),[3] and accusations against India's media for its double standard in reporting on Islamist terror but not violence by the Hindu majority against India's minorities. The emails also explain specific stratagems used by IM. For example, the emails explained that the Delhi bombing was intended to highlight the fact that even high-security zones are not safe and that Bengaluru was targeted because it is India's IT capital (Naqvi 2008).

It is possible that IM issues its manifestos, as opposed to simple claims of responsibility knowing that it has further attacks in the works, thus adding credence and force to its message.

4.3 Attacks Against Public Sites and Claims of Responsibility

IM frequently makes claims of responsibility for terrorist attacks. In some cases, these claims can be viewed as extremely credible as they were made either shortly before or in the midst of a short-term (under 1 h) attack. For instance:

[3] For an in-depth discussion of IM's grievances regarding the BJP, see Chap. 2.

- IM announced its existence with a claim of responsibility for a series of three, near-simultaneous attacks on courtrooms in three separate cities on November 23, 2007.[4]
- Later, IM claimed responsibility by email for the series of blasts in markets in Jaipur in May 2008.
- On July 26, 2008, a series of attacks in Ahmedabad in Gujarat were claimed by IM in an email sent five minutes prior to the first blast, lending credibility to their claim of responsibility.
- Likewise, on September 13, 2008, IM sent an email from a hacked account at 6:27 pm, claiming responsibility for a series of explosions at markets that commenced at roughly 5:55 pm. A few days later, Indian law enforcement raided an IM safe-house at Batla House in Delhi. The evidence obtained by Indian law enforcement lends further credibility to the likelihood that IM carried out the attacks.
- On September 20, 2010, on the anniversary of the aforementioned Batla House encounter, IM claimed responsibility for a shooting that targeted tourists and a failed car bomb near the Jama Masjid in Delhi.

TP-Rule PS-2. IM attacks public sites two months after months in which:

- IM has been communicating about its campaign.

Support = 4
Probability = 100%, *Inverse Probability* = 100%,
Negative Probability = 0%

IM has a long track record of claiming responsibility under its own name.

TP-rule (PS-3) below shows the connection between claims of responsibility and *future* attacks carried out by IM.

TP-Rule PS-3. IM attacks civilians two months after months in which:

- IM made claims of responsibility for certain terrorist attacks.

Support = 4
Probability = 100%, *Inverse Probability* = 100%, *Negative Probability* = 0%

[4] This attack was coded as targeting a government site, not a public site, but is included here as an example of IM's issuing claims of responsibility after its high profile attacks.

4.4 Attacks on Public Sites and Conferences Held by Indian Mujahideen

IM periodically holds secret conferences in order to plan attacks. While the Internet has proven a boon to terrorist operations and the ability to manage far-flung operations, terrorist groups still find that in-person meetings are necessary to make major decisions and plan operations. IM is by no means unique in this regard. Other terrorist groups have also needed to augment online communications with face-to-face meetings (Hunt 2006; Trujillo 2005). For IM, a series of meetings starting with a January 2008 meeting in the jungles in Pavagadh in the western Indian state of Gujarat were necessary to plan the string of bombings with which IM terrified India throughout the year 2008 (TNN 2008).

TP-Rule PS-4. IM attacks public sites four month after months in which:

- IM held a conference.

Support = 4
Probability = 100%, *Inverse Probability* = 100%,
Negative Probability = 0%

As in the case of the previous rules, TP-rule (PS-4) above specifies conditions under which IM attacked a public site within 4 months of holding a conference.

The above TP-rule shows a close connection between conferences and IM attacks on public sites. Given TP-rules (PS-1) through (PS-3), it seems that IM holds a conference, engages in additional rhetoric to drum up support for an attack, prepares logistically for an attack, and finally follows through on the attack. This sequence of action however is speculative and would need data on the internal operations of IM to which the authors are not privy.

4.5 Attacks on Public Sites and Membership in Other Non-State Armed Groups

Islamist groups in South Asia are connected through a web of ties that enable them to leverage each other's expertise. IM members like Riyaz Bhatkal have sought refuge in Pakistan after escaping the reach of Indian law enforcement.

Membership in other NSAGs allows IM operatives to train in camps run by those disposed positively towards IM, as well as to receive logistics support for their operations. Shared membership with other NSAGs strengthens IM's capabilities and operational reach.

There are innumerable examples of IM operatives with extensive links to Pakistani-backed terrorist groups. One prominent example includes one of IM's co-founders Abdul Subhan Qureshi, who had trained with LeT (Tankel 2011) and was a key organizer of the Ahmedabad blasts (Gupta 2011). Tadyyantavide Nasir, the IM operative accused of masterminding the Bengaluru blasts of 2008, had links to IM, its predecessor SIMI, and to LeT dating back to 2000 (Madhusoodan 2009).

TP-rule (PS-5) below states that when information about IM members belonging to another NSAG emerges, IM is likely to carry out attacks against public sites.

The same rule holds true with a 3-month offset with a slightly diminished support of four, and with a lower support of three for 5-month offsets (Though rules with such low support fell below this study's threshold for consideration, we mention it here because of its relevance to the rules that did have support of four or more). Figure 4.3 shows the situation graphically for the period since February 2007.

TP-Rule PS-5. IM attacks public sites two month after months in which:

- IM shared members with other non-state armed groups.

Support = 5
Probability = 100%, *Inverse Probability* = 100%, *Negative Probability* = 0%

4.6 Attacks on Public Sites and the Entertainment of Diplomatic Relations Between India and Pakistan

It has been hypothesized that when the Indian and Pakistani governments are nearing a diplomatic breakthrough, forces interested in disrupting peace arrangements, such as elements of the Pakistani military and/or Pakistani Islamist radicals, derail negotiations by unleashing their proxies to carry out devastating terrorist attacks. This hypothesis was put forward by numerous individuals following the deadly November 26, 2008 terrorist attacks in Mumbai that were subsequently proved to be the work of LeT (Subrahmanian et al. 2012; Tankel 2011). Moreover, the testimony of David Coleman Headley, an American citizen of Pakistani origin, in a Chicago courtroom (Rotella 2011) further validated the hypothesis that Pakistan's ISI both planned the attacks and was intimately involved in its operational execution.

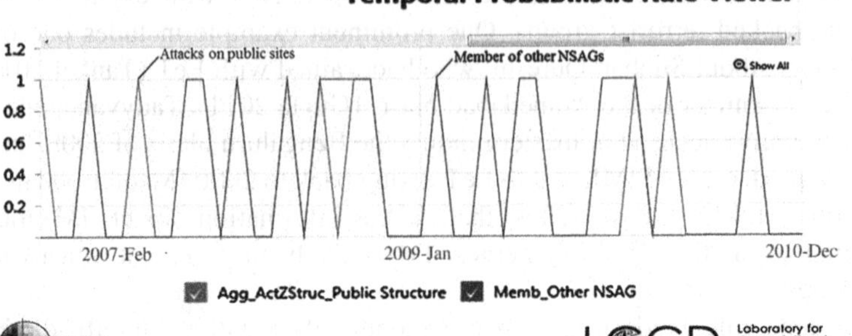

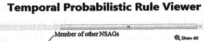

Fig. 4.3 Times when IM operatives were mentioned in media sources as members of other non-state armed groups preceding IM attacks on public structure

However, past studies of LeT were unable to substantiate a consistent link between efforts to derail the diplomatic process and any single kind of LeT attack.[5]

In the case of IM, however, the study did derive a TP-rule that substantiated a link between warming of diplomatic ties between India and Pakistan and attacks on public sites by IM. This is shown as TP-rule (PS-6).

IM's rhetoric does not discuss Indian-Pakistani relations. There is evidence nonetheless that relations between the rivals were warming in 2008, including unconfirmed reports of substantial back-channel talks over Kashmir. These talks were effectively sidelined by the Mumbai attack in November 2008 (Coll 2009). It is certainly plausible that LeT and other Pakistani terror groups expanded their support for IM during periods of warming Indo-Pakistani relations, even if that conflict was not central to IM's own grievances.

TP-Rule PS-6. IM attacks public sites five months after months in which:

India and Pakistan entertain diplomatic links.

Support = 4
Probability = 100%, *Inverse Probability* = 100%, *Negative Probability* = 0%

Figure 4.4 below illustrates this situation in greater detail, graphically visualizing the link between warming of diplomatic relations between India and Pakistan and subsequent attacks on public sites by IM since September 2008.

[5] Such a link may exist when we consider LeT attacks in totality as opposed to LeT attacks of one type at a time, which is what was done previously (Subrahmanian et al. 2012).

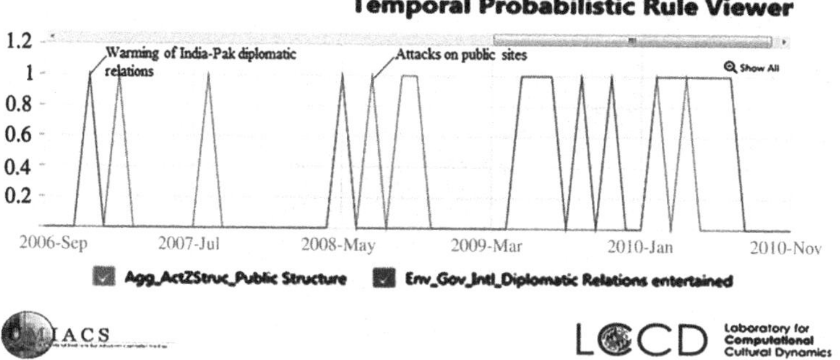

Fig. 4.4 IM attacks public sites a few months after India and Pakistani relations improve

4.7 Operational Hypothesis

The timing of the conditions that trigger attacks on public sites by IM and the actual attacks themselves generate some interesting hypotheses. Figure 4.5 shows this situation.

This figure shows that, for each independent variable shown in Fig. 4.5, when the condition captured by the independent value is true, there is a high probability of an IM attack on a public site within a few months.

This suggests a hypothesis about operational planning by IM members. Perhaps when there is a distinct prospect of warm diplomatic ties between New Delhi and Pakistan's elected government (as opposed to the military), elements in Pakistan's military worry about loss of influence and set in motion a chain of events that culminates in an IM conference a month later, leading to an operational decision to carry out attacks (though targeting may not be determined at such a conference).

Though IM cadres are routinely believed to have joint membership in other NSAGs, the coding method we used only flags them as members when there is explicit evidence (e.g., news reporting that IM founder Riyaz Bhatkal joined LeT after he fled to Pakistan, see Ali 2012). As a consequence, the independent variable in the IM dataset would only be set when there are explicit reports of IM's interaction with other NSAGs. Perhaps the interactions between IM and other NSAGs become more pronounced during periods of planning an attack, leading to this interaction being flagged more frequently.

At this point, IM may also issue a statement about its intentions and strategy, usually in the context of a claim of responsibility for an attack.

Perhaps it is the process of such mobilization that leads to arrests of IM personnel 1–2 months before the planned IM operations.

Perhaps emboldened by the energy and strength of its recruits, and possibly to energize them even further, IM claims responsibility for other attacks occurring at that time, whether orchestrated by them or not.

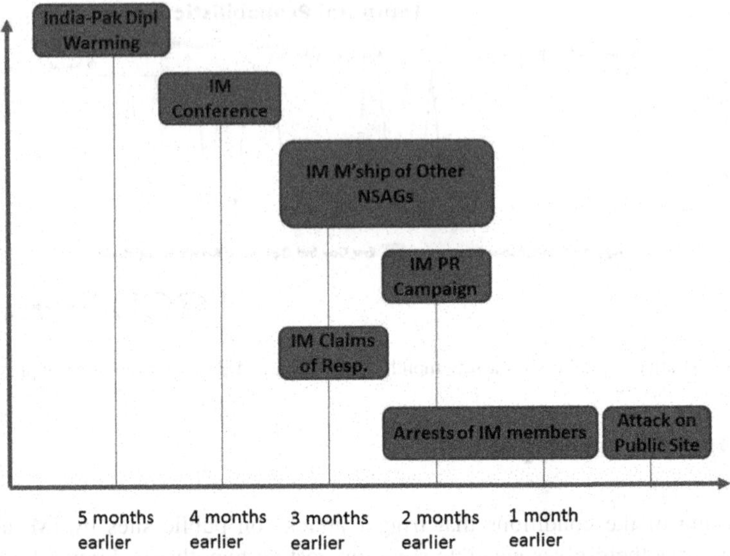

Fig. 4.5 Time line showing conditions true in months leading up to an IM attack on public sites

And finally, with their recruits energized, logistical plans in place, and with public statements about upcoming attacks, IM moves forward and launches an attack on a public place.

Though Sect. 4.7 is consistent with the observed facts and the data provide evidence for the relationships between the environmental variables and IM's propensity to attack public sites, it should be noted that the sequence of events in Fig. 4.5 is hypothetical. It cannot be verified without access to the internal deliberations and planning processes within IM and their relationship to their sponsors, the Pakistani ISI.

4.8 Conclusions and Policy Options

This chapter has attempted to identify conditions that precede attacks launched by IM on public sites. Simply put, it has identified the following conditions.

- IM attacks on public sites seem to be preceded by five months by a warming of diplomatic relations between India and PakistanPakistan.
- IM attacks on public sites seem to be preceded by four months by conferences held by IM.
- IM attacks on public sites seem to be preceded by three months by increased evidence of interaction between IM and other Islamist non-state armed groups.

- Additionally, three months before IM attacks on public sites, IM appears to start claiming responsibility for other attacks (irrespective of the type of entity targeted).
- Two months before IM attacks on public sites, IM issues statements about its campaign, intentions, and strategy.
- For 1–2 months before IM attacks on public sites, there seems to be an uptick in arrests of IM personnel—in general, between one and 21 IM personnel are arrested.

These observations are not necessarily causative—rather, they represent events that co-occur with a 1–5 month delay. Some of these observations however may still serve as "canaries in the coalmine", helping Indian and U.S. law enforcement and counter-terrorism officials determine terrorist threat levels, increase surveillance of known IM operatives, and allocate investigative resources.

Clearly, any warming of Indo-Pakistani ties is an immediate warning that greater vigilance is necessary. And conferences of IM personnel would be yet another sign of potential trouble. If these are accompanied or followed in short order by increased reports of interaction between IM personnel and other NSAGs, by IM statements regarding its operational campaign, and increased levels of claims of responsibility, then the levels of caution around public sites, particularly soft targets like markets, need to be increased, especially at times when the markets are likely to be crowded with shoppers.

While not all variables listed above are easily observable, some certainly are (e.g., when the organization discusses its terror campaigns in its manifestos, increased tendency to make claims of responsibility for other attacks). Intelligence sources may have access to some of the other variables as well (e.g. through individuals who have infiltrated the organization).

References

Ali SA (2012) Cops heard of Riyaz Bhatkal for the first time. The Times of India February 11. http://articles.timesofindia.indiatimes.com/2012-02-11/mumbai/31049240_1_twin-blasts-riyaz-bhatkal-india-and-zaveri-bazaar

Coll S (2009) The back channel: India and Pakistan's secret Kashmir talks. The New Yorker March 2. http://www.newyorker.com/reporting/2009/03/02/090302fa_fact_coll

Fair C (2010) Students Islamic Movement of India and the Indian Mujahideen: an assessment,.Asia Policy January

Fernandes J (2012) Journalist arrested in Bangalore for links with Indian Mujahideen: sources. NDTV August 30 http://www.ndtv.com/article/south/journalist-arrested-in-bangalore-for-links-with-indian-mujahideen-sources-260793

Gupta S (2011) Indian Mujahideen: the enemy within. Hachette, Gurgaon

Hunt E (2006) Virtual incompetence. The Weekly Standard August 18 http://www.weeklystandard.com/Content/Public/Articles/000/000/012/594icnci.asp

Madhusoodan (2009) Tatiyantavide Nasir's terror Inc was based out of Yeshwantpur. Daily News and Analysis, December 8.http://www.dnaindia.com/bangalore/report_tadiyantavide-nasir-s-terror-inc-was-based-out-of-yeshwantpur_1321337

Naqvi S (2008) Not a pang of guilt. Outlook India September 29 http://www.outlookindia.com/article.aspx?238507

Ramesh R (2008) Indian Mujahideen claims responsibility for Jaipur blasts,. The Guardian May 15 http://www.guardian.co.uk/world/2008/may/15/india

Rediff.com (2009) A chronology of the 2003 Mumbai twin blasts case. July 27 http://news.rediff.com/report/2009/jul/27/a-chronology-of-the-2003-mumbai-twin-blasts-case.htm

Rotella S (2011) Pakistan and the Mumbai attacks: the untold story. ProPublica

Roul A (2010) After Pune, details emerge on the Karachi project and its threat to India. CTC Sentinel April 3 http://www.ctc.usma.edu/posts/after-pune-details-emerge-on-the-karachi-project-and-its-threat-to-india

Subrahmanian VS, Mannes A, Shakarian J, Sliva A, Dickerson J (2012) Computational analysis of terrorist groups: Lashkar-e-Taiba. Springer, New York

Swami P (2008) Pakistan and the Lashkar's jihad in India. The Hindu December 9 http://www.hindu.com/2008/12/09/stories/2008120955670800.htm

Swami P (2008) Not just a claim, a manifesto for jihad. The Hindu May 17 http://www.hindu.com/2008/05/17/stories/2008051754761100.htm

Tankel S (2011) Storming the world stage: the story of Lashkar-e-Taiba. C. Hurst & Co, London

TNN (2008) Indian Mujahideen, another face of SIMI. Times of India August 17 http://articles.timesofindia.indiatimes.com/2008-08-17/ahmedabad/27929682_1_simi-members-serial-blasts-safdar-nagori

Tripathi R (2008) Serial blasts rock Delhi; 30 dead, 90 injured. Times of India September 14 http://timesofindia.indiatimes.com/Cities/Delhi/Delhi_30_dead_as_5_bombs_go_off_in_45_min/rssarticleshow/3479914.cms

Trujillo H (2005) The radical environmentalist movement,. In Aptitude for destruction volume 2: case studies of organizational learning in five terrorist groups. (Brian Jackson ed) The RAND Corporation, Santa Monica

Chapter 5
Bombings

Abstract The Indian Mujahideen (IM) has carried out numerous bombings, including those at crowded city markets, transportation hubs, and commercial districts. This chapter discusses IM's primary mode of attack, bombs planted at targets. This chapter shows close connections between bombings carried out by IM and prior arrests of IM operatives, IM communications about their operational campaign, membership of IM operatives in other groups such as SIMI, LeT and HuJI, as well as diplomatic initiatives between India's government and the government of Pakistan. In addition, structural conditions such as Hindu-Muslim tensions within India play an important role in predicting bombings by IM.

Bombings have been Indian Mujahideen IM's preferred weapon of terror. Below is a partial listing of attacks in which IM is a suspect. In addition to the attacks listed below, the previous chapter on attacks on public sites lists the series of IM bombings across India that terrorized the country through 2008:

- On February 23, 2005, IM was responsible for an attack at a Hindu holy bathing place (known as a ghat) in the Hindu holy city of Varanasi. The attack was believed to have been caused by two improvised explosive devices built using pressure cookers used all over India to cook rice, similar to those used in the 2013 Boston Marathon bombings (Gupta 2011). Nine people were killed and over 23 injured in the 2005 Varanasi attack.
- On July 28, 2005, IM, possibly under the auspices of its predecessor organization SIMI, was believed to be responsible for the attack on the Shramjeevi Express that runs between Rajgir in Bihar (in the eastern part of India) and Delhi. Thirteen people were killed and 52 injured in the attack, which was originally blamed on leaking gas cylinders before an IM and SIMI link was established (Gupta 2011).[1]
- On March 7, 2006, 16–28 people died and 95–101 were injured in two simultaneous bombings in Varanasi, one near the Sankat Mochan temple and the

[1] Another possible suspect is HuJI, an Islamist group with origins in Pakistan from the early 1980s during the Soviet occupation of Afghanistan.

V. S. Subrahmanian et al., *Indian Mujahideen*, Terrorism, Security, and Computation, DOI: 10.1007/978-3-319-02818-7_5,

other near a train station. Several other bombs did not detonate. Though the attack was originally claimed as an attack by Lashkar-e-Qahar (Associated Press 2009), it was revealed in 2008 that this was an IM operation (Gupta 2011).

- In July 11, 2006, IM is believed to have been the principal party responsible for seven simultaneous bombings on first-class compartments during Mumbai's commuter rush hour. In what is perhaps the worst attack carried out by IM to date, 209 people were killed and 714 injured. This is one of many IM attacks where ascribing responsibility is difficult. Many experts agree that IM was largely responsible, although IM may have worked closely with and received operational guidance from LeT.[2] The arrest of two IM operatives, Mujeeb Shaikh and Abu Faisal (TNN 2011), appear to confirm IM's involvement.
- In October 13, 2008, between 10 and 18 bombs were planted in Guwahati, the capital of Assam, with 83 dead and 300 and 400 injured. The bombs included three car bombs. Though responsibility was claimed by HuJI and the Assamese separatist group United Liberation Front of Assam (ULFA) as well as IM, IM is a key suspect.
- More recently, on Feb 13, 2010, a bomb went off in a crowded bakery in the western Indian city of Pune, killing 17 people and injuring over 54. The attack was believed to be a joint IM and LeT operation.

IM has been responsible for numerous bombings. This chapter discusses several variables that are closely related to these bombings.

- *Government Arrests of IM Personnel.* As in the case of Chap. 4, attacks on public sites, the findings indicated that when IM members are arrested, bombings seem to follow within 1–3 months with high probability. The same behavior was also true for LeT (Subrahmanian et al. 2012).
- *Communications about their Campaigns by IM.* Likewise, when IM is discussing their terror campaign in their communications, bombings tend to follow within 2, 4, or 5 months. This factor also played a role in IM's attacks on public sites (Chap. 4) and is relevant to attacks that IM carried out jointly with other terrorist groups.
- *Conferences organized by IM.* As in the case of IM attacks on public sites detailed in Chap. 4, there is also a link between conferences organized by IM and bombings 4 months later. This lends credibility to the hypothesis that IM uses such conferences for operational decision making and/or planning methods, perhaps selecting targets and/or allocating personnel to plan and execute such attacks.
- *Membership in Other NSAGs.* As in the case of attacks on public sites (Chap. 4), there is a close relationship between months when IM members also belong to other NSAGs and subsequent land bombings perpetrated by IM two, three, or five months later. Though this precise variable did not play a role with respect to attacks carried out by IM's close ally, LeT, (Subrahmanian et al. 2012) points

[2] And, of course, there are many who believe that IM is just an arm of LeT.

out that when LeT was providing support to other Islamist groups, they (LeT) were move likely to carry out terrorist attacks a few months later.

- *Diplomatic Relations between India and Pakistan.* As in the case of attacks by IM on public sites (Chap. 4), there is a close relationship between the warming of diplomatic relations between India and Pakistan and bombings three to five months later by IM.
- *Internal Conflict.* Lastly, the algorithm came up with an interesting rule, noting that when India is undergoing deadly internal conflicts, IM appears highly likely to carry out attacks 1 month later. Many internal conflicts in India involve Hindu-Muslim riots and it is possible that such riots inflame IM passions and strengthen its resolve to carry out deadly attacks to justify their self-declared role as a defender of the Islamic population in India.

IM's bombings appear to be carried out under circumstances very similar to their attacks on public sites. One reason could be that a number of events were coded as both bombings and as attacks on public sites (e.g., the aforementioned 2011 Pune bakery bombing). Another reason could be that IM selects soft targets when it decides to launch attacks and is neutral as to whether the targets are bus stations, trains, markets, or the outside of stadiums, as long as the potential to create maximal bloodshed and mayhem remains high.

5.1 Bombings and Arrests of Indian Mujahideen Personnel

As in the case of attacks on public sites, there seems to be a strong relationship between arrests of IM personnel and bombings carried out by IM a few months later. TP-rule (BOMB-1) below shows one such rule.

TP-Rule BOMB-1. IM carries out bombings two months after months in which:

- Between 1 and 21 IM members were arrested.

Support = 5
Probability = 100%, *Inverse Probability* = 100%,
Negative Probability = 0%

The above TP-rule is also valid with a support of 4 when either a 1-month or a 3-month offset is considered instead of 2 months. Thus, this rule is validated in a number of ways.

Figure 5.1 shows a graph plotting bombings carried out by IM and prior months when IM members were arrested.

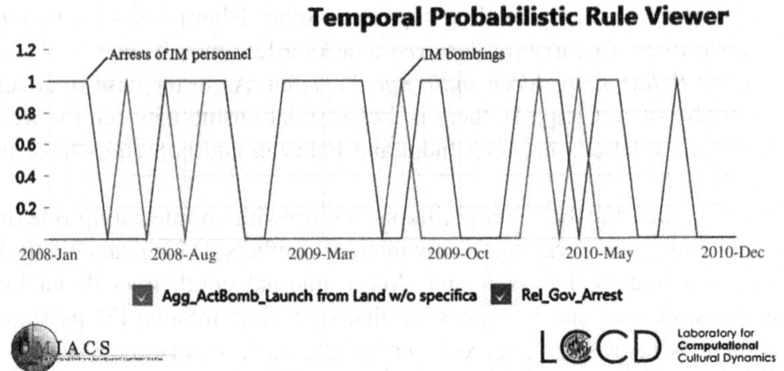

Fig. 5.1 Relationship between IM carrying out bombings and prior arrests of IM personnel

After high-profile bombings, Indian police carry out mass arrests of individuals suspected of links to IM. However, top tier commanders often evade these sweeps, which sometimes arrest lower level operatives or people who are unfortunate enough to have links to IM members. In July 2008, Indian police arrested 21 suspected IM members, but IM continued to carry out deadly attacks that year as top leaders such as the Bhatkals evaded capture (NDTV 2010).

5.2 Bombings and Indian Mujahideen Communications About Their Terror Campaign

As in the case of IM attacks on public sites (cf. Chap. 4), months in which IM communications discuss the strategy and operations of their terror campaign preceded bombings by a period of 2, 4, and 5 months. IM tends to email manifestos just before, after, or during its bombings. As discussed in Chap. 4, these manifestos provide details as to the organization's motivations and grievances. For example, the emailed manifesto after the May 2008 serial bombings in Jaipur threatened to target India's tourism industry and argued that if Muslims are not safe in India, then "infidels" would also not be safe (Swami 2008). Because IM's attacks tend to follow within a few months of one another, the manifestos are also issued months before the next attack. It is possible that IM manifestos could provide clues about the next attack.

TP-Rule (BOMB-2) below describes the situation when a 2-month offset holds.

TP-Rule BOMB-2. IM carries out bombings two months after months in which:

- IM issued communications about its terror campaign.

Support = 5
Probability = 100%, *Inverse Probability* = 100%,
Negative Probability = 0%

When a time offset of either 4 or 5 months is considered, a TP-rule is generated that is identical to (BOMB-2) except that the support drops down slightly to 4.

Additionally, IM periodically carries out joint bombings with other groups like LeT and SIMI. TP-rule (BOMB-3) is similar to the one listed above and addresses the case of joint bombings carried out by IM.

TP-Rule BOMB-3. IM carries out joint bombings two months after months in which:

- IM issues communications about its terror campaign and
- IM shares members with other non-state armed groups.

Support = 4
Probability = 100%, *Inverse Probability* = 100%,
Negative Probability = 0%

Figure 5.2 shows the situation associated with the above rule as well as the link between bombings and IM members also being members of other non-state armed groups.

The content of IM's communications may serve as a leading indicator of attacks to come. It would be prudent for appropriate security agencies in the U.S. and India to carefully study IM statements for indications of the group's strategy and use the findings as guides to increase surveillance of IM, improve security around potential targets, and increase internal danger and threat advisories similar to the Department of Homeland Security's National Terrorism Advisory System in the U.S.

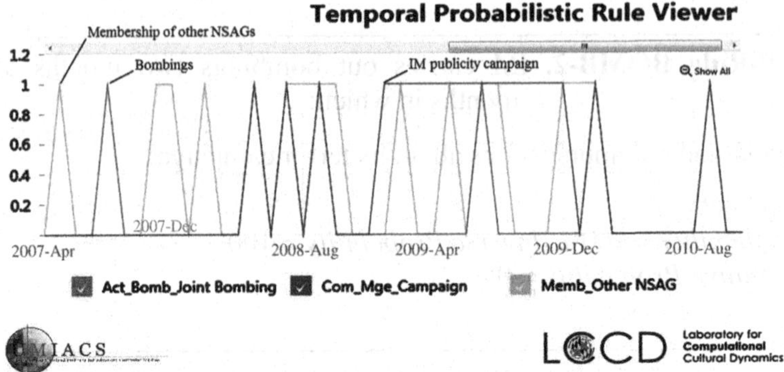

Fig. 5.2 Joint bombings carried out by IM in conjunction with another organization occurring a few months after IM issued communications about its campaign and its shared members with other non-state armed groups

5.3 Bombings and Conferences Held by Indian Mujahideen

Another rule indicates that when IM holds conferences, it is extremely likely that attacks by IM follow 4 months later. A similar rule also applies to the case of attacks on public sites carried out by IM.

As discussed in Chap. 4, as a clandestine group, IM needs to gather its key operatives in order to decide on and plan its attacks. The 2008 bombing campaign was devised in a series of meetings in early 2008 (TNN 2008).

TP-Rule (BOMB-4) below shows the relationship between holding conferences and subsequent bombings by IM.

TP-Rule BOMB-4. IM carries out bombings four months after months in which:

- IM held a conference.

Support = 4
Probability = 100%, *Inverse Probability* = 100%,
Negative Probability = 0%

As in the case of IM's public discussions of its terror campaign, the fact that IM is holding a conference can be viewed almost as a declaration of intent by them to carry out bombings a few months later.

Figure 5.3 below is a visual depiction of TP-rule (BOMB-4), showing how meetings between high-level IM operatives are followed by bombings.

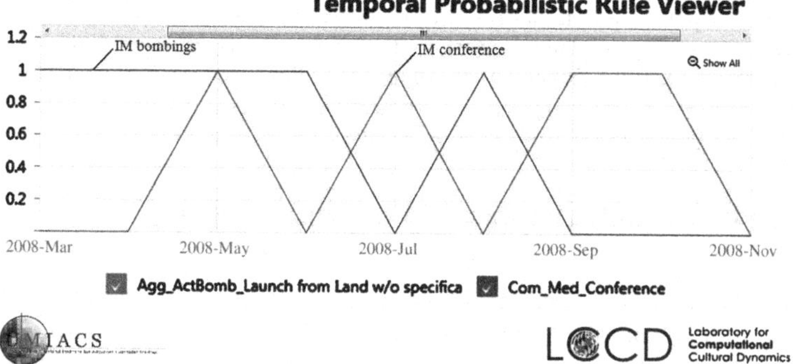

Fig. 5.3 Relationship between IM holding a conference and IM bombings a few months after

5.4 Bombings and Claims of Responsibility for Past Attacks

IM has a long history of claiming responsibility for its attacks, although in some cases other groups also claim responsibility for the same attack. TP-Rule (BOMB-5) below shows that 2 months after those months in which IM claimed responsibility for an attack, IM was likely to carry out another bombing.

> **TP-Rule BOMB-5.** IM carries out bombings in which 0-69 people are killed two months after months in which:
>
> - IM claimed responsibility for an attack.
>
> *Support* = 4
> *Probability* = 100%, *Inverse Probability* = 100%,
> *Negative Probability* = 0%

Explicit claims of responsibility for past attacks by IM may be a harbinger of future bombings in the offing. Perhaps IM feels that its bold claims for past terrorist attacks energizes its base, motivating the IM rank and file for the next attack. The lengthy claims of responsibility and manifestos include both justifications and threats. In November 2007, after simultaneous bombings of courthouses in Varanasi, Lucknow, and Faizabad, IM formally announced its existence with an emailed manifesto claiming responsibility for the attack. IM specifically blamed the courts for not bringing justice to India's Muslims:

> The Supreme Court, the high courts, the lower courts and all the Commissions have utterly failed to play an impartial role regarding Muslim issues. Narendra Modi who presided

> over the 2002 massacres of Muslims in Gujarat is given a clean chit whereas the victims still run from pillar to post for justice. Even the 92 Mumbai culprits roam freely and enjoy Government security. All the anti-Muslim pre-planned riots, arson, rapes, losses of lives and properties are still awaiting justice. The list is endless! (Raman 2011, para. 14).

5.5 Bombings and Membership in Other NSAGs

IM has a long history of connections to other armed groups. Long believed to be the more militant and hardline outgrowth of SIMI (Fair 2010), IM also is viewed by many experts as a proxy within India—a kind of fifth column—for both the Pakistani terror group LeT and their sponsors, Pakistan's ISI.

It therefore could be argued that the condition of sharing members with other NSAGs is a relatively constant condition for IM, a condition that is almost always true. In the dataset however, IM was only coded as sharing members with other NSAGs if one of several news sources *reported* it. It was *not* recorded as being true for a given month if there was not an explicit report saying that this was the case. Reports of IM cadres being members of other NSAGs may be more likely when this happens in increasing numbers. Perhaps such reports occur in the press when IM operatives are moving more freely amongst groups, drawing notice from security officials.

TP-Rule (BOMB-6) below shows that when bombings by IM killed zero to 69 people, it was preceded by months when IM's members also belonged to other NSAGs.[3]

> **TP-Rule BOMB-6.** IM carries out bombings two months after months in which:
>
> - IM members were also members of other NSAGs.
>
> *Support* = 5
> *Probability* = 100%, *Inverse Probability* = 100%,
> *Negative Probability* = 0%

The same TP-rule is also valid with a slightly reduced support of 4 when both a 3-month and 5-month offset are considered.

A related rule pertains to *joint bombings* carried out by IM. For instance, the July 2006 a Mumbai train bombing that wreaked havoc and killed over 200 commuters is widely believed to have been executed by IM cadres, but with

[3] It should be noted that the range of deaths of 0–69 includes bombings in which no one was killed, not months in which no bombings occurred.

involvement from both LeT and SIMI. The February 2010 bombing of the German Bakery at Pune that killed 17 may have been another joint IM-LeT operation, possibly carried out under the auspices of LeT's "Karachi Project" (Roul 2010).

TP-rule (BOMB-7) below shows that when IM members are also members of other NSAGs, they are likely to carry out joint bombings 2 months later.

TP-Rule BOMB-7. IM carries out bombings two months after months in which:

- IM members were also members of other NSAGs.

Support = 6
Probability = 100%, *Inverse Probability* = 100%,
Negative Probability = 0%

5.6 Bombings and Warming of Diplomatic Relations Between India and Pakistan

Another rule states that when there is a thaw in diplomatic relations between India and Pakistan, IM is likely to carry out bombings 3 or 5 months later. There have been numerous points at which Pakistani-Indian relationships have improved. In July 2005, India and Pakistan were engaged in talks (BBC News 2005); in October 2005, Delhi was struck by a pair of blasts at a market during Eid-ul-Fitr and Diwali. The blasts are thought to have been the work of IM, probably in conjunction with LeT (Gupta 2011). Nonetheless, the talks continued. In March 2006, Varanasi was struck by two near-simultaneous blasts (Gupta 2011). Despite these attacks by IM and LeT, talks continued. In December 2006, former President Musharraf stated that Pakistan and India were very close to an agreement and blamed LeT for the continual breakdowns of Indian-Pakistani talks (The News 2010).[4]

This pattern has continued to the present. In September 2009, Indian and Pakistani officials met on the sidelines of the United Nations General Assembly. Although they did not formally re-start talks, to meet at all less than 1 year after the Mumbai assault was remarkable (Polgreen and Mekhennet 2009). 5 months later, IM bombed the German Bakery at Pune.

It is possible that IM was working in conjunction with LeT to stymie peace negotiations. Alternately, IM may have been doing this on its own initiative or was

---

[4] Many noted scholars however, including former CIA operative Bruce Reidel (2013), assert that Musharraf was playing a double game, using LeT as a proxy force to carry out his goals and then blaming them for whatever went wrong.

carrying out attacks for its own reasons. A third possibility is that IM carried out these attacks at the behest of Pakistan's military establishment which did not want warmer diplomatic relations between India and Pakistan. A fourth possibility is that LeT, particularly in periods when it was forced to curtail its activities in Jammu and Kashmir, expanded its support for IM, but did not give IM specific operational direction.

TP-Rule (BOMB-8) shows this rule for 3-month offsets.

TP-Rule BOMB-8. IM carries out bombings five months after months in which:

- There was a warming of diplomatic relations between India and Pakistan.

Support = 5
Probability = 100%, *Inverse Probability* = 100%,
Negative Probability = 0%

A much weaker rule with support of 3 also holds when a 3-month rather than 5-month offset is considered. The lower support might indicate that IM is occasionally able to launch attacks within 3 months of improved diplomatic ties between Delhi and Islamabad, but 5 months appears to be the more likely situation.

5.7 Bombings and Internal Conflict Within India

An interesting TP-rule about IM indicates that when there is internal violence within India, bombings are likely to follow 1 month later.

TP-Rule BOMB-9. IM carries out bombings one month after months in which:

- There was internal violence within India.

Support = 4
Probability = 100%, *Inverse Probability* = 100%,
Negative Probability = 0%

In fact, a variant of this rule says that when the internal conflict led to between 2 and 15 individual casualties in a given month, there were likely to be IM bombings 1 month later.

TP-Rule BOMB-10. IM carries out bombings one month after months in which:

- There was internal violence within India that killed 2-15 individuals.

Support = 4
Probability = 100%, *Inverse Probability* = 100%,
Negative Probability = 0%

5.8 Policy Options and Conclusions

Figure 5.4 shows a time line of the variables associated with bombings carried out by IM.

Figure 5.4 shows that IM carries out bombings 5 months after diplomatic ties between India and Pakistan (and in a few cases the United States) show signs of getting stronger.

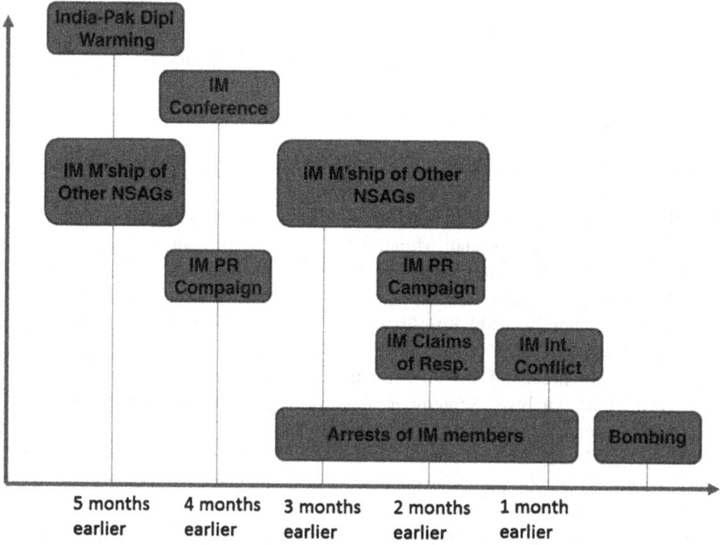

Fig. 5.4 Time line showing bombings carried out by IM and conditions that are true prior to the bombing

1 month after the diplomatic ties are announced and 4 months before the bombing, IM holds a conference and issues statements regarding its terror campaign.

3 months before the bombing, IM's shared membership with other NSAGs is also active, suggesting that IM may be seeking the support or advice of their more experienced terrorist colleagues in groups like LeT, and possibly the Pakistani intelligence agencies (though we do not have independent evidence of this).

2 months prior to the bombing, IM operatives are arrested (it is not clear why this is the case), possibly further inflaming IM and firing up their publicity campaign.

Internal violence within India also appears correlated with IM attacks. IM has referred to discrimination by Hindus against minority groups within India besides Muslims, and it is possible that this violence adds fuel to IM's ideological fire. India however is a vast country with a number of ongoing internal conflicts, and there may not be a relationship between the overall level of internal violence within India and IM attacks.

As in the case of Chap. 4 on attacks on public sites, we recommend that in order to mitigate bombings, a close eye should be kept on IM's activities whenever diplomatic ties between India and Pakistan seem to be on the mend. Moreover, IM conferences need to be carefully monitored by security agencies and when IM issues claims of responsibility or discusses publicly its terror campaign, IM may be planning bombings.

References

Associated Press (2009) Bombay mourns bombing victims. February 11. http://www.cbsnews.com/2100-202_162-1815512.html.

BBC News (2005) Hindus protest at Ayodhya attack., July 6 http://news.bbc.co.uk/2/hi/south_asia/4654593.stm.

Fair C (2010) Students Islamic Movement of India and the Indian Mujahideen: an assessment, Asia policy, January.

Gupta S (2011) Indian Mujahideen: the enemy within. Hachette, Gurgaon.

NDTV (2010) Pune blast: The Indian Mujahideen connection? February 15. http://www.ndtv.com/article/cities/pune-blast-the-indian-mujahideen-connection-16293.

Polgreen L, Mekhennet S (2009) Militant network is intact long after Mumbai siege. The New York Times September 29, 2009. http://www.nytimes.com/2009/09/30/world/asia/30mumbai.html?_r=1&ref=world.

Raman B (2011) past threats of Indian Mujahideen (IM) against judiciary analysis. Eurasia Review News and Analysis September 7. http://www.eurasiareview.com/07092011-past-threats-of-indian-mujahideen-im-against-judiciary-analysis/.

Roul A (2010) After Pune, details emerge on the Karachi Project and its threat to India. CTC Sentinel April 3 http://www.ctc.usma.edu/posts/after-pune-details-emerge-on-the-karachi-project-and-its-threat-to-india.

Subrahmanian VS, Mannes A, Shakarian J, Sliva A, Dickerson J (2012) Computational analysis of terrorist groups: Lashkar-e-Taiba. Springer, New York.

Swami P (2008) Not just a claim, a manifesto for jihad, The Hindu May 17 http://www.hindu.com/2008/05/17/stories/2008051754761100.htm.

The News (2010) JUD not a soft underbelly: Musharraf, The News. October 10 http://www.thenews.com.pk/latest-news/2638.htm.

TNN (2008) Indian Mujahideen, another face of SIMI., Times of India August 17 http://articles.timesofindia.indiatimes.com/2008-08-17/ahmedabad/27929682_1_simi-members-serial-blasts-safdar-nagori.

TNN (2011) Mumbai blasts: Indian Mujahideen duo in focus, fears grow of bigger strike in Gujarat., Times of India July 16 http://articles.timesofindia.indiatimes.com/2011-07-16/india/29781629_1_im-operatives-im-module-terror-attacks.

Chapter 6
Simultaneous and Timed Attacks

Abstract IM has specialized in carrying out simultaneous or consecutive attacks where multiple geographically dispersed targets, usually but not always in the same city, are attacked simultaneously or in quickly timed succession. These attacks are often on soft targets like city markets. This chapter examines the conditions under which IM carries out such attacks.

Over the years, IM has carried out numerous simultaneous attacks in which multiple targets are attacked during a relatively short period of time. Attacks are either simultaneous or timed to occur in quick succession. Such multi-pronged attacks require substantial technical and operational sophistication and can be extremely deadly. Most of IM's simultaneous attacks have been bombings, and even when the bombs are poorly designed, these attacks spread fear and terror. Examples of IM's multi-pronged attacks include:

- In August 2012, five bombs were planted at various locations in Pune. Fortunately the explosions did not take any lives, although one injury was reported (Daily News and Analysis 2012).
- In July 2011, three bombs were detonated in Mumbai in the Opera House district, the Zaveri Bazaar, and the central district of Dadar. This time, the attacks killed 26 and injured 141 (PTI 2011).
- In October 2008, three car bombs were set off in Guwahati, the capital of the Eastern state of Assam; 83 people were killed and between 300 and 400 people were injured. Though IM claimed responsibility, HuJI is also suspected. While HuJI, LeT, and JeM are all groups with similar motivations (Hussain 2007), HuJI is closely allied with the Taliban. In addition to HuJI, the separatist group ULFA was also suspected in this attack (Kalita 2008).
- One month earlier, in September 2008, IM was held responsible for five attacks that killed between 26 and 30 people and injured between 100 and 130 people in Delhi. All these blasts occurred in areas with high population densities. IM claimed responsibility for these attacks within minutes of the attacks occurring, leading to a high presumption of responsibility on their part. IM's emailed claim

V. S. Subrahmanian et al., *Indian Mujahideen*, Terrorism, Security, and Computation, DOI: 10.1007/978-3-319-02818-7_6,

of responsibility was sent via a hacked computer account, highlighting the technical sophistication of IM operatives (Yadav and Bhatia 2008).

- About a month and a half earlier, in July 2008, IM was responsible for 20 low intensity bombs in Ahmedabad, all within a period of about 45 min. Moreover, as families of the dead and injured gathered at the City Trauma Center about 30 min later, a car bomb exploded by the hospital, leading to further death and destruction. Indian security forces located and defused another two car bombs and an additional 23 smaller bombs. All told, 57 were killed and over 100 were wounded. IM claimed responsibility for these attacks a few minutes *prior* to the start of the attacks, leading to a high probability that they were, in fact, responsible for the attacks. The email was sent using unsecured Internet accounts over a Wi-Fi channel (Swami 2008a, b).
- The day before the Ahmedabad attacks, on July 25, 2008, the city of Bengaluru was struck by eight low intensity blasts within a 30-min period. Though LeT (Subrahmanian et al. 2012; Tankel 2011; John 2012) claimed responsibility, Indian intelligence suspected IM actually executed the attack because of the similarity in design to the bombs used in the Ahmedabad attacks the next day (Swami 2008a, b). There was also an anonymous caller to a local television station who claimed it was an IM operation (Times of India 2008). Later police investigations in this attack highlighted the links between LeT and IM (Nanjappa 2009). The attack claimed two lives and injured seven to eight people.
- In May 2008, the Rajasthani city of Jaipur was savaged by a series of nine bombs in seven locations, all over the course of roughly 30 min. The simultaneous attacks led to 60–80 deaths and 100–216 injuries. Though IM claimed responsibility for the attacks via an email, Indian police also listed HuJI as a possible perpetrator (PVTR 2008).
- In August 2007, within a quarter of an hour, a series of bombs targeted lawyers in the cities of Varanasi, Faizabad, and Lucknow. Several of the bombs were planted in bicycles. Between 15 and 18 were killed in these attacks that also injured 57–81 people. Five minutes before the bombings, IM announced itself to the world through a manifesto emailed to the media throughout India (Gupta 2011).
- In August 2007, two bombs were simultaneously detonated in another of India's IT hubs, the city of Hyderabad, killing 42 people and injuring 54. Furthermore, police defused two additional bombs. Suspicion again fell on IM, though some analysts also believe that HuJI might have been the perpetrator or that IM and HuJI executed the attack jointly (Gupta 2011).
- In May 2007, the city of Gorakhpur in Uttar Pradesh was targeted with three bombs planted in milk jugs, all at a single market. One of the jugs was placed in a market, the other at the exit to the market, and the third at an electrical transformer. The plan may have been for the first jug in the midst of the market to explode, causing panicked shoppers to head to the exits where a second bomb was waiting to claim further victims. Targeting the electric transformer may have been a strategy to induce further panic when the lights went out. Fortunately, the bombs failed to detonate (Gupta 2011).

- In July 2006, seven bombs exploded within first class compartments on Mumbai's suburban commuter trains. The train bombings killed 209 people and injured 714, making it one of the bloodiest terrorist attacks on Indian soil. There is disagreement as to which organization carried out the attack. Many suspect this was a joint operation and there is reason to believe that IM planted the bombs on the trains (Gupta 2011). Lashkar-e-Qahhar, a shadowy outfit closely connected with LeT, claimed responsibility, asserting that 16 people were involved in carrying out the attacks and that all 16 were safe (Times of India 2006; Business Standard 2006). An email to an Indian TV channel allegedly sent from the central Indian city of Indore (Outlook India 2006) claimed that the attacks were in retaliation for India's actions in Kashmir. The communiqué further asserted that the train bombings were a part of a series of attacks planned on major tourist attractions.
- Earlier in March 2006, two simultaneous bombs went off in the holy city of Varanasi, one near the Sankat Mochan temple and another at the railway station. A third bomb was discovered prior to detonation. Though the attack was initially claimed by Lashkar-e-Qahhar in connection with the July 2006 train bombings, this operation was reportedly an IM operation (Gupta 2011). The attacks claimed 16–28 lives and injured 95–101 people.
- In May 2005, two pressure-cooker based IEDs were planted at the holiest bathing spot on the Ganges in Varanasi. Though the attack was originally believed to be a pressure cooker explosion, in 2006, Indian police declared this as a terrorist incident. Nine people were killed and over 20 injured in this attack (Gupta 2011).
- In August 2003, IM's predecessor organization, SIMI, was held responsible for planting explosives in two taxis that exploded at the Gateway of India and Zaveri Bazaar in Mumbai. The attack was said to be a response to the massacres of Muslims in Gujarat. Fifty-two people died and 184 were injured. SIMI and LeT are believed to have carried out this attack in a joint operation (Ali 2012).

This lengthy enumeration of attacks highlights the fact that simultaneous or consecutive attacks on multiple targets are a favored IM tactic and that the vast majority of IM attacks are of this type.

The remaining sections of this chapter describe temporal probabilistic (TP) rules discovered by the Stochastic Temporal Analysis of Terrorist Events (STATE) system. The TP-rules capture the conditions under which the Indian Mujahideen carry out simultaneous attacks.

6.1 Simultaneous and Timed Attacks and Conferences

Chapters 4 and 5 already hypothesize the existence of strong connections between the times when IM holds conferences and attacks on public sites (Chap. 4) and bombings (Chap. 5). Rule (SA-1) below provides evidence that four months after an IM conference, there is a very high probability of simultaneous attacks occurring.

> **TP-Rule SA-1**. IM carries out simultaneous attacks 4 months after months in which:
>
> - IM organized a conference.
>
> *Support* = 4 *Probability* = 100 %, *Inverse Probability* = 100 %.

Though this rule has only moderate support of 4, IM has not been in existence as long as groups like Lashkar-e-Taiba for which TP-rules were required to have higher support for them to be considered useful (Subrahmanian et al. 2012).

> **TP-Rule SA-2**. IM carries out simultaneous attacks 4 months after months in which:
>
> - IM organized a conference.
>
> *Support* = 4 *Probability* = 100 %, *Inverse Probability* = 100 %.

Figure 6.1 displays a sample of our data showing that IM carries out consecutive attacks a few months after holding a conference.

A similar TP-rule was derived for the variable associated with "timed attacks" where attacks were carefully timed. This variable is also closely related to the "simultaneous attack" variable.

The algorithm also derived some weaker TP-rules that provide additional support to the hypothesis that IM carries out consecutive attacks a few months after conferences.

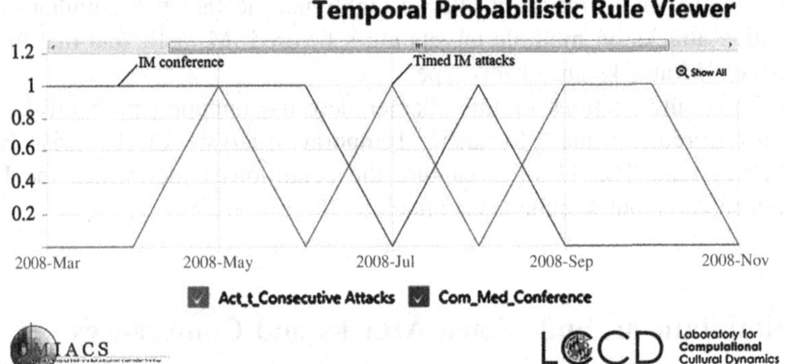

Fig. 6.1 Relationship between IM holding a conference and IM carrying out consecutive attacks a few months later

- The STATE system discovered another TP-rule that has virtually the same properties as TP-rule (SA-1) above, but with a slightly weaker support of 3. This TP-rule states that one month after an IM conference, there is a 100 % probability (and 100 % inverse probability) of consecutive attacks by IM.
- An identical TP-rule to that in the previous bullet was also discovered with respect to a two-month offset. This TP-rule also had a support of 3.

As discussed in Chaps. 4 and 5, a clandestine group such as IM needs to gather its key operatives in order to decide on and plan its attacks. The 2008 bombing campaign was devised in a series of meetings in early 2008.

6.2 Simultaneous and Timed Attacks by IM and Arrests of IM Personnel

Rules discussed in previous chapters also noted that timed attacks are correlated with arrests of IM personnel. In particular, timed attacks seem to follow within 2 months of arrests of IM personnel as shown in TP-rule (SA-3) below.

TP-Rule SA-3. IM carries out timed attacks 2 months after months in which:

- Between 1 and 21 IM operatives were arrested.

Support = 4 *Probability* = 100 %, *Inverse Probability* = 100 %.

If we allow a weaker support of 3, then STATE also found some TP-rules that connected simultaneous attacks with arrests of IM personnel. Though these rules will not be used in detail for the policy generation chapter, they support our findings that arrests of IM personnel are followed a few months later by attacks on public sites (TP-rule PS-1 in Sect. 4.2) and by bombings (TP-rule BOMB-1 shown in Sect. 5.1).

This TP-rule only has a support of 3 (very low) but it provides additional evidence that arrests of IM personnel are followed by timed attacks a few months later.

TP-Rule SA-4. IM carries out simultaneous attacks 2 months after months in which:

- Between 1 and 21 IM operatives were arrested.

Support = 3 *Probability* = 100 %, *Inverse Probability* = 100 %.

When the support level required for TP-rules is dropped to 2, there is further evidence for this hypothesis. A TP-rule identical to (SA-2) applies to a 3-month offset with a weakened support of 2; moreover, when looking at a one-month offset, STATE derived a TP-rule stating that when 12–21 IM personnel were arrested in a given month, there is a 100 % probability of simultaneous attacks 1 month later.

6.3 Simultaneous and Timed Attacks and Membership in Other Armed Groups

We also found evidence that when IM operatives are also members of other NSAGs, we can expect timed attacks a few months later.

IM operatives have a long history of membership in other NSAGs, including LeT and HuJI. Overlapping membership in part reflects the amorphous structure and overlapping ideologies of these organizations. As disaffected Indian Muslims drifted towards extremism, they came into contact with Pakistani terrorist groups such as LeT.

At the same time, LeT was seeking Indian recruits to expand its ability to strike India. The relationship between LeT and IM is complex and cannot simply be described master-proxy. LeT's support and training however was crucial to IM's establishment, and Indian Muslims who received LeT training brought essential skills and resources to IM.

The investigation into the 2008 Bengaluru bombings highlights one example of how individuals with overlapping memberships in terrorist groups can facilitate operations. Sarfaraz Nawaz had been a member of IM's predecessor organization SIMI. Living in Oman, he met Ali Abdul Aziz al-Hooti, a wealthy entrepreneur and LeT financier. When Tadiyantavide Nasir, an Islamist political activist who helped inspire Nawaz needed funds for a terror cell in Kerala, Nawaz facilitated the transfer of funds, brought Nasir into contact with al-Hooti, and helped arrange for Nasir's recruits to get LeT training in Jammu and Kashmir. Nasir's cell went on to carry out the 2008 Bengaluru bombing (Swami 2009).

TP-Rule SA-5. IM carries out timed attacks 2 months after months in which:

- IM members were members of other non-state armed groups.

Support = 4 *Probability* = 100 %, *Inverse Probability* = 100 %.

6.4 Conclusions and Policy Options

This chapter identified conditions preceding simultaneous attacks and timed attacks launched by IM on multiple targets. Figure 6.2 summarizes the TP-rules about simultaneous and/or timed attacks derived by our STATE system.

The STATE system supports the proposition that 4 months after IM holds a conference, IM can be expected to launch simultaneous/timed attacks. This adds

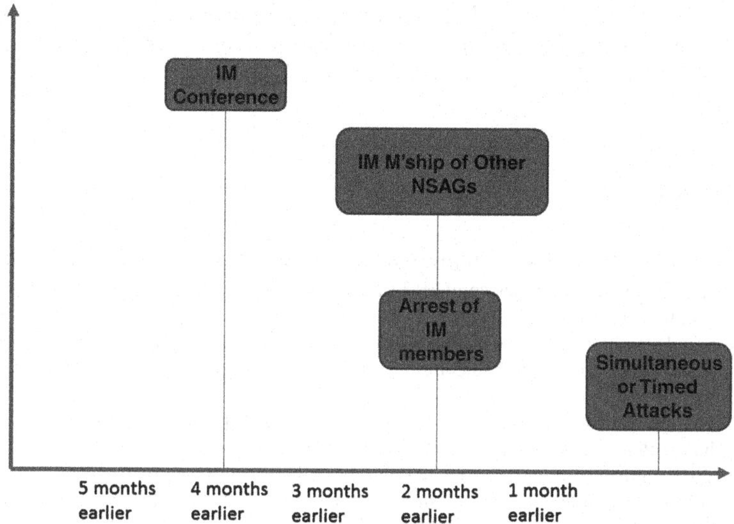

Fig. 6.2 Relationship between timing of events in IM's environment and simultaneous and/or timed attacks carried out by IM

support to the TP-rules that IM conferences are followed a few months later by attacks and that these conferences likely serve either a planning need and/or involve making operational decisions regarding which attacks to move forward within the subsequent few months.

Figure 6.2 echoes similar figures in other chapters describing events prior to attacks. However, in the case of simultaneous and consecutive attacks, the events preceding the operation relate to IM planning operations, not IM's communications activities or broader events external to IM such as Indian-Pakistani diplomatic relations. The variables are IM conferences where leaders meet to plan attacks, IM members linking with other non-state armed groups which provides critical logistical support for operations, and the arrests of IM members which may reflect increased IM activity prior to an attack. This highlights that IM's complex and carefully timed multiple bombings require extensive planning and support from other terrorist groups.

Moreover, STATE also derived TP-rules that reinforce the TP-rules presented in Chaps. 4 and 5, asserting a link between arrests of IM personnel and subsequent attacks on public sites and bombings. This chapter has shown that there is evidence to support the hypothesis that two months after IM personnel are arrested, IM is likely to carry out timed attacks as well. In general, the results of this chapter and the preceding two chapters imply that arrests of IM personnel are followed by subsequent IM attacks of different types.

This is not to imply that arrests of IM personnel be stopped—there is no suggestion of a causal link between arrests of IM personnel and subsequent simultaneous attacks (or in fact attacks on public sites or bombings). Our findings are predictive, i.e., arrests of IM personnel increases the likelihood of IM attacks of different types in the succeeding months. Consequently, a greater degree of preparedness and alertness after arrests of IM personnel would be in order.

References

Ali SA (2012) Cops heard of Riyaz Bhatkal for the first time. Times of India February 11. http://articles.timesofindia.indiatimes.com/2012-02-11/mumbai/31049240_1_twin-blasts-riyaz-bhatkal-india-and-zaveri-bazaar.

Business Standard (2006) Mumbai blasts: we did it: Lashkar-e-Qahhar. Press Trust of India July 16. http://www.business-standard.com/india/news/mumbai-blasts-we-did-it-lashkar-e-qahhar/3025/on.

Daily News and Analysis (2012) Pune serial blasts: as it happened. August 1. http://www.dnaindia.com/india/commentary_pune-serial-blasts-as-it-happened_1722867.

Gupta S (2011) Indian Mujahideen: the enemy within. Hachette, Gurgaon.

Hussain Z (2007) Frontline Pakistan: the struggle with militant Islam. Columbia University Press, New York.

John W (2012) The Caliphate's soldiers: the Lashkar-e-Tayyeba's long war. Amaryllis

Kalita P (2008) Car bombs were used in Guwahati. Times of India November 1. http://articles.timesofindia.indiatimes.com/2008-11-01/india/27895609_1_guwahati-blasts-khagen-sarma-ulfa.

Nanjappa V (2009) Revealed: LeT's close ties with Indian Mujahideen. February 7. http://www.rediff.com///news/2009/feb/07beng-let-close-ties-with-indian-mujahideen.htm.

Outlook India (2006) Email from Lashkar-e-Qahhar came from Indore: Police. July 18. http://news.outlookindia.com/items.aspx?artid=400006.

PTI (2011) Mumbai blasts: death toll rises to 26. July 20. http://www.hindustantimes.com/India-news/Mumbai/Mumbai-blasts-Death-toll-rises-to-26/Article1-727292.aspx.

PVTR (2008) Serial bomb blasts in Jaipur. International Centre for Political Violence and Terrorism Research, A Centre of the S. Rajaratnam School of International Studies. http://www.pvtr.org/pdf/RegionalAnalysis/SouthAsia/Serial%20bomb%20blasts%20in%20Jaipur.pdf.

Subrahmanian VS, Mannes A, Shakarian J, Sliva A, Dickerson J (2012) Computational analysis of terrorist groups: Lashkar-e-Taiba. Springer, New York.

Swami P (2008) 'Indian Mujihideen' claims responsibility. The Hindu July 27 http://www.hindu.com/2008/07/27/stories/2008072759280100.htm.

Swami P (2008) Not just a claim, a manifesto for jihad. The Hindu May 17 http://www.hindu.com/2008/05/17/stories/2008051754761100.htm.

Swami P (2009) The Indian Mujahidin and Lashkar-i-Tayyiba's transnational networks. CTC Sentinel Jun 15. http://www.ctc.usma.edu/posts/the-indian-mujahidin-and-lashkar-i-tayyiba%E2%80%99s-transnational-networks.

Tankel S (2011) Storming the world stage: the story of Lashkar-e-Taiba. C. Hurst & Co, London.

Times of India (2006) Lashkar-e-Qahhar claims responsibility for 11/7. Times of India July 16. http://articles.timesofindia.indiatimes.com/2006-07-16/india/27822270_1_vile-parle-station-blasts-suburban-trains.

Times of India (2008) Indian Mujahideen, another face of SIMI. Times of India August 17. http://articles.timesofindia.indiatimes.com/2008-08-17/ahmedabad/27929682_1_simi-members-serial-blasts-safdar-nagori.

Yadav S, Bhatia R (2008) 5 blasts in Delhi, 25 dead: Indian Mujahideen claims responsibility. Tribune News Service September 13http://www.tribuneindia.com/2008/20080914/main1.htm.

Chapter 7
Total Deaths in Indian Mujahideen Attacks

Abstract As described in previous chapters, the Indian Mujahideen have carried out numerous attacks against public sites, bombings, and consecutive/timed attacks. This chapter investigates the number of people killed in IM attacks rather than the specific type of attack itself. The findings of this chapter identify conditions that can be monitored on IM's environment that are predictive of the number of people killed in IM attacks.

Figure 7.1 below shows the number of attacks carried out by Indian Mujahideen IM on an annual basis over the last 7 years. There was a steady increase in the number of attacks from 2005–2008, none in 2009, and an increase in 2010. This was followed by a drop in IM attacks in 2011.

On the other hand, Fig. 7.2 shows the total numbers killed by IM during the above period.

Figure 7.2 shows two bars for each year corresponding to a lower bound and an upper bound, respectively, of the number of people killed by IM in that year. IM killed the most people in 2006 (225–237 killed) and 2008 (209–249). No killings were recorded in 2009, followed by a slightly increase in 2010 (19 killings) and 2011 (27 killings).

The change after 2008 may be a function of two factors, one being a crackdown on LeT after its November 2008 assault on Mumbai, thereby being less able to provide support to IM. At the same time, arrests of IM operatives began to affect IM's operational capabilities. One particular event, the September 2008 Batla House encounter in New Delhi, may have had a significant effect. A week after the 2008 Delhi serial bombings, police raided an IM safe house where they killed two IM members, captured another, and obtained a wealth of documents about the organization that led to further arrests (Gupta 2011; Roul 2009). These arrests were a serious blow to IM's capabilities.

Finally, Fig. 7.3 shows the number of people injured in IM attacks, excluding fatalities.

V. S. Subrahmanian et al., *Indian Mujahideen*, Terrorism, Security, and Computation,
DOI: 10.1007/978-3-319-02818-7_7,

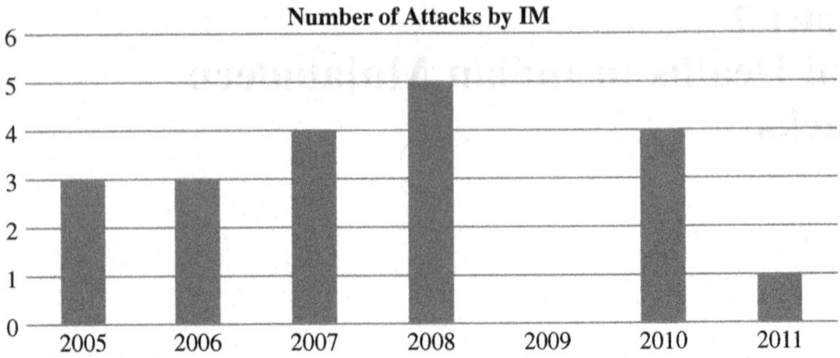

Fig. 7.1 Numbers of attacks by IM since 2005

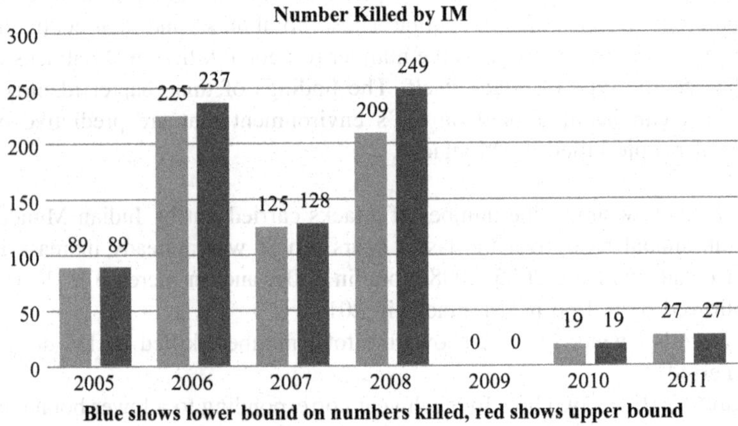

Fig. 7.2 Total numbers of people killed by IM. *Blue* shows lower bound on numbers killed, *red* shows upper bound

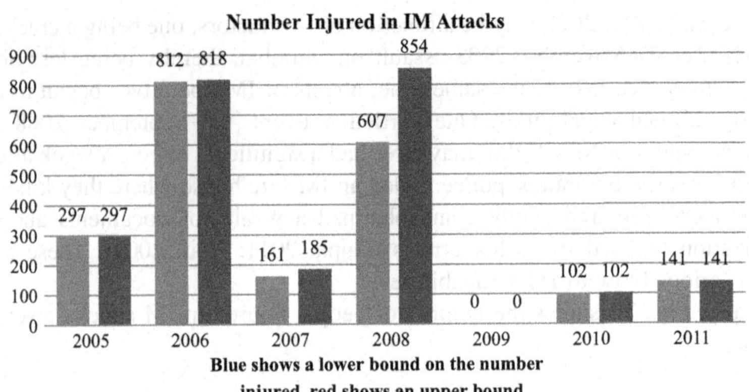

Fig. 7.3 Numbers injured in IM attacks, 2005–2011, excluding fatalities. *Blue* shows a lower bound on the number injured, *red* shows an upper bound

According to Fig. 7.3, the largest numbers of injuries occurred in 2006 and 2008. There were no injuries in 2009, followed by an increase in 2010 and 2011 (albeit significantly lower than 2006 and 2008). Most of the injuries in 2006 were caused by the Mumbai train blasts (714 injured). In contrast, in 2008, injuries were a function of numerous and smaller attacks in geographically dispersed cities, including Jaipur, Bengaluru, Ahmedabad, New Delhi, and Guwahati.

In the remaining sections, we describe rules discovered by our Stochastic Temporal Analysis of Terrorist Events (STATE) computer system that capture the conditions under which the I.M carry out attacks, usually, but not always leading to fatalities.

7.1 Number Killed and Conferences

In previous chapters, we showed a strong association between IM holding conferences and IM attacks on public sites, bombings, and simultaneous/timed attacks. TP-rule (TK-1) below provides evidence that 4 months after an IM conference, there is a very high probability of fatalities, further supporting the hypothesis that conferences held by IM are strong predictors of attacks to come.

TP-Rule TK-1. IM carries out one or more attacks that kill between 0 and 69 people four months after months in which:

- IM organized a conference.

Support = 4

Probability = 100%, *Inverse Probability* = 100%.

This TP-rule needs to be read carefully. Even though the rule seems to state that 0 people could have been killed, the rule also makes clear that at least one attack was attempted and that the number of fatalities in such attacks was between 0 and 69.

7.2 Numbers Killed and Communications About its Terror Campaign

The STATE system also automatically generated TP-rules showing that there is a strong relationship between IM's statements about its ability to carry out further attacks and subsequent attacks leading to fatalities conducted a few months later.

TP-Rule TK-2. IM carries out one or more attacks that kill between 0 and 69 people two months after months in which:

- IM communicates about its ability to carry out attacks.

Support = 3

Probability = 100%, *Inverse Probability* = 100%.

TP-rule (TK-2) has slightly weaker support (of 3) than we would like; nonetheless, it is significant because it relies on IM's own statements about its capabilities. In the manifestos issued just before and after major attacks, IM often presents the operational rationale for its attacks. These are not threats per se, but rather statements about the organization's intentions and capabilities. For example after the May 2008 Jaipur attacks, the organization issued an email entitled "Indian Mujahideen's Declaration of Open War Against India". The email stated that the attacks were specifically intended to target tourism and religious sites (Swami 2008).

7.3 Number Killed and Arrests of Indian Mujahideen Personnel

We have already seen in previous chapters that arrests of IM personnel are followed a few months later by attacks on public sites, bombings, and simultaneous/timed attacks. Similarly, arrests of IM personnel increase the probability of IM-sponsored attacks that kill people, as shown in TP-rule (TK-3) below.

TP-Rule TK-3. IM carries out one or more attacks that kill between 0 and 69 people one month after months in which:

- Between 1 and 21 IM personnel were arrested by the government.

Support = 4

Probability = 100%, *Inverse Probability* = 100%.

In fact, an identical rule holds when we consider both a 2 and 3-month offset (though for the latter, support drops to 3). We thus conclude that arrests of IM personnel are followed by attacks that cause fatalities 1, 2, or 3 months later.

As in previous chapters, we emphasize that this does not imply that Indian police should stop arresting IM personnel—especially those apprehended during the commission of a crime or during efforts to plan a crime. Rather, we merely note that arrests do not have a deterrent effect on IM. Instead, they appear to spur IM to show their rank and file that they are capable of carrying out attacks *despite* these arrests, thus energizing their core supporters. It is also possible that in periods such as 2008, when multiple IM plans were already set, a few arrests were insufficient to disrupt ongoing operations. Over time however, the cumulative arrests do seem to impact the organization. After the arrests of 2008, IM was not active in 2009.

7.4 Numbers Killed and Improving Diplomatic Relations

We have already noted that IM carries out both attacks on public sites and bombings a few months after warming of diplomatic relations between India and Pakistan. Not surprisingly, this phenomenon also applies to attacks that cause death and injuries.

The evidence for this is slightly weaker however than in the case of TP-rules (TK-1), (TK-2), and (TK-3).

TP-Rule TK-4. IM carries out one or more attacks that kill between 0 and 69 people five months after months in which:

- There is a warming of diplomatic relations between India and Pakistan.

Support = 3
Probability = 100%, *Inverse Probability* = 100%.

TP-rule (TK-4) is made more credible since an identical rule applies using a time offset of 3 months rather than five. This rule is particularly interesting because although IM's primary motivation seems to be issues within India, it receives support from the Pakistani terrorist group LeT, which has acted in its own right to disrupt Pakistani-Indian relations. It is possible that LeT support for IM increases in periods when relations between the South Asian rivals are warming, such as in 2008, a year in which LeT's assault on Mumbai froze the slow improvement of Pakistani-Indian relations (Coll 2009).

7.5 Numbers Killed and Relationship with Other NSAGs

Membership in other NSAGs is closely related to subsequent attacks by IM on public sites and bombings, as well as simultaneous/timed attacks. It is not surprising that membership in other NSAGs is closely related to fatalities. As discussed in previous chapters, overlapping memberships facilitate the transfer of money and expertise from more experienced Pakistani terrorist groups to IM.

TP-Rule TK-5. IM carries out one or more attacks that kill between 0 and 69 people two months after months in which:

- IM's members were reported as belonging to other non-state armed groups.

Support = 5
Probability = 100%, *Inverse Probability* = 100%.

TP-rule (TK-5) has strong support. We also note that though some believe that IM members have always overlapped with other NSAGs (e.g., HuJI and LeT), we only coded this property of belonging to other NSAGs when it was explicitly noted in the news and other sources it.

TP-rule (TK-5) is further supported by an almost identical rule that holds for a 3-month offset, with a slightly lower support of 4. Moreover, with a slightly lower support of 3, this rule also holds for a 5-month offset.

7.6 Numbers Killed and Claims of Responsibility

Prior claims of responsibility made by IM also relates to the number killed in new attacks. Specifically, this rule says that when IM makes claims of responsibility for an attack, we can expect *new* attacks 2 months later that lead to 0 to 69 casualties. TP-rule (TK-6) below shows this situation.

TP-Rule TK-6. IM carries out one or more attacks that kill between 0 and 69 people two months after months in which:

- IM claimed responsibility for an attack.

Support = 4
Probability = 100%, *Inverse Probability* = 100%.

IM has announced its presence through emailed manifestos, often just before attacks occur or while the attacks are in progress, that provide substantial evidence of their responsibility. For example the email sent just before the May 2008 Jaipur bombings included pictures of the bicycles used in the bombings (Swami 2008). In another instance, shortly before the Ahmedabad bombings, on July 26, 2008

Gujarat police received an email stating, “We the Indian Mujahideen have carried out attacks and will continue to do so. Stop us if you can” (Nanjappa 2008). These claims were followed by attacks 2 months later. After a period of relative quiet in 2009, IM struck again with a bombing in Pune in February 2010. The claim of responsibility was also followed 2 months after the Pune bombings by a series of bombings in Bengaluru.

7.7 Numbers Killed and Internal Conflict Within India

The STATE system also found an association between occurrence of *internal conflict* within India and subsequent occurrence of IM attacks that resulted in casualties, as shown in TP-rule (TK-7).

TP-Rule TK-7. IM carries out one or more attacks that kill between 0 and 69 people one month after months in which:

- There was internal conflict within India with individual casualties.

Support = 4
Probability = 100%, *Inverse Probability* = 100%.

India experiences frequent internal conflicts at a relatively low level (e.g. riots, clashes on the basis of ethnicity). When these clashes involve attacks on Hindus by Muslims or conversely by attacks on Muslims by Hindus, tensions are inflamed.

The above rule TK-7 states that 1 month after internal conflicts leading to casualties, IM would likely carry out attacks resulting in fatalities.

TP-Rule TK-8. IM carries out one or more attacks that kill between 0 and 69 people one month after months in which:

- Internal conflict within India led to between 2 and 15 fatalities.

Support = 4
Probability = 100%, *Inverse Probability* = 100%.

Rule (TK-7) is further supported by a more detailed rule also with a 1-month offset, TP-rule (TK-8). Both rules indicate that internal conflict leading to fatalities precedes IM attacks 1 month later.

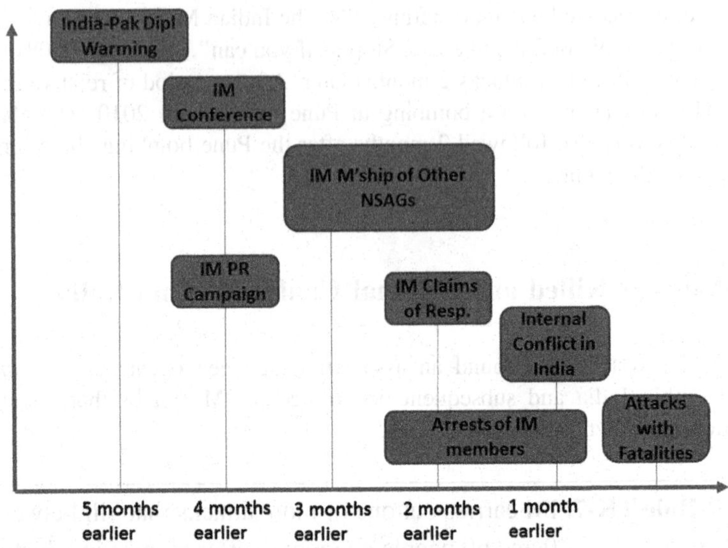

Fig. 7.4 Events preceding IM attacks with fatalities

It should be noted that the STATE system also found a TP-rule with a 5-month offset that is almost identical to TP-rule (TK-7) except that the rule has a slightly weaker support of 3.

Figure 7.4 presents a timeline showing the relationship between attacks resulting in fatalities and preceding variables predicting the likelihood of an attack.

One to two months after a significant breakthrough in bilateral Indo-Pakistani ties, there is a significant likelihood of IM convening a conference, together with movement of personnel between IM and NSAGs (e.g., LeT, HuJI). Perhaps an operational decision to execute attacks has already been taken and the purpose of the shuffle of personnel between IM and its allied armed jihadist groups is to set in motion the selection of personnel needed to execute such attacks. Perhaps the purpose of the conference is for target selection and to plan the logistics required for carrying out such attacks. The STATE system allowed us to conclude that 4 months after IM holds such a conference, we expect IM to launch simultaneous attacks. Moreover, at roughly the same time, IM issues statements regarding its intentions regarding its terror campaign.

A month later (and 2 months before attacks that claim casualties), IM starts claiming responsibility for other attacks launched. In addition to promulgating their message to the Indian public, these manifestos show their rank and file that IM has successfully executed attacks, thus energizing their recruits with enthusiasm for *jihad*.

One to two months prior to the attack, IM operatives end up being arrested. Whether this is due to carelessness on their part or due to extreme diligence on the part of Indian intelligence and security services is difficult to say. However, it is likely that increased operational planning and communications are flagged by the Indian security establishment, leading to arrests. We note that we also derived TP-rules in Chaps. 4 and 5 asserting a link between arrests of IM personnel and subsequent attacks on public sites and bombings. Chapter 6 provided somewhat weaker evidence to support the hypothesis that 2 months after IM personnel are arrested, IM is likely to carry out simultaneous attacks as well. In general, the results of this chapter and the preceding three chapters suggest that arrests of IM personnel are followed by subsequent IM attacks of different types.

We are not suggesting that arrests of IM personnel be stopped, and we are not aware of a causal link between arrests of IM personnel and subsequent attacks. What we are suggesting is that when IM personnel are arrested, the threat of attacks of different types in the succeeding months greater than when no IM personnel are arrested. This suggests that a greater degree of preparedness and alertness after arrests of IM personnel would be in order. IM's behavior of carrying out retaliatory attacks after arrests is similar to LeT's behavior (Subrahmanian et al. 2012).

In other words, when there are signs of significant improvement in relations between India and Pakistan, it is necessary for the Indian security establishment to increase vigilance and surveillance of terrorist networks. Furthermore, when the improved diplomatic climate is followed in short order by a conference of IM operatives, such vigilance and surveillance should be significantly enhanced as attacks that result in civilian deaths are almost certain to follow.

Tactically, this suggests that detailed electronic surveillance of all known IM targets should be quickly and significantly increased when a conference occurs in a climate of thawing India-Pakistan relationships. Intelligence assets within IM should be maximally exploited (within the bounds of the law) to secure additional information on what is brewing.

References

Coll S (2009) The back channel: India and Pakistan's secret Kashmir talks. The New Yorker, March 2. http://www.newyorker.com/reporting/2009/03/02/090302fa_fact_coll.

Gupta S (2011) Indian Mujahideen: the enemy within. Hachette, Gurgaon.

Nanjappa V (2008) Minutes before blasts email said: 'Stop us if you can'. Rediff.com July 26. http://www.rediff.com/news/2008/jul/26ahd3.htm.

Roul A (2009) India's home-grown jihadi threat: a profile of the Indian Mujahideen. Jamestown Foundation Terrorism Monitor March 3. http://www.jamestown.org/single/?no_cache=1&tx_ttnews[tt_news]=34577&tx_ttnews[backPid]=7&cHash=7d49c63104.

Subrahmanian VS, Mannes A, Shakarian J, Sliva A, Dickerson J (2012) Computational analysis of terrorist groups: Lashkar-e-Taiba. Springer, New York.

Swami P (2008) Pakistan and the Lashkar's jihad in India. The Hindu December 9. http://www.hindu.com/2008/12/09/stories/2008120955670800.htm.

Swami P (2008) Not just a claim, a manifesto for jihad. The Hindu May 17. http://www.hindu.com/2008/05/17/stories/2008051754761100.htm.

Chapter 8
Computing Policy Options

Abstract This chapter describes the methodology and the algorithm used to automatically generate policy options. It provides a mathematical definition of a policy against IM. The chapter presents an algorithm to compute all policies (in accordance with the mathematical definition of policy) that have high probability of significantly reducing all types of attacks carried out by IM (except for attacks on holidays). We were able to find one such policy. This policy will be discussed in detail in Chap. 9.

Chapters 4–7 of this book describe the conditions under which IM carries out four types of terrorist attacks or attempted attacks. The variables related to these attacks are partially summarized in the figures shown at the end of each of these chapters. Although these summary figures do suggest possible policies to mitigate one type of attack, they do so at the risk of aggravating another kind of attack. Deriving "good" policies requires a process of simultaneously considering how the wide variety of attacks carried out by IM might be mitigated.

In this chapter, we briefly describe the computational methods used to generate policies automatically from the TP-rules described and discussed in Chaps. 4–7. Appendix B summarizes the set of all TP-rules described in this book. Our policy computation algorithms build upon the methods to use linear programming to compute all minimal models of logic programs developed by one of the authors in a series of papers (Bell et al. 1994a, b).

A slight variant of this chapter was first published in (Subrahmanian et al. 2012).

V. S. Subrahmanian et al., *Indian Mujahideen*, Terrorism, Security, and Computation, DOI: 10.1007/978-3-319-02818-7_8,

8.1 Policy Analysis Methodology

In this section, we briefly describe the methodology we used to automatically generate policies that aim to significantly reduce IM's violent activities. Our methodology involves the steps outlined below.[1]

- *Step 1 Time Offset and Probability Elimination:* As all rules in this book have time offsets of one, two, three, four, or five months, we eliminated all the time offsets and probabilities from the TP-rules. The resulting set of "reduced rules" all have the form

$$A(i) \leftarrow B_1 \& \ldots. \& B_n$$

  where $A(i)$ describes an action taken by LeT with intensity level i and B_1 & .... & B_n is a conjunction of environmental literals (atoms or negated atoms). Intuitively, this rule can be read now as: IM will take action A with intensity level i (with high confidence) within one to five months when B_1 & .... & B_n is true.
- *Step 2 Policy Computation:* Using the remaining rules, we developed methods to compute the set of all policies (i.e., ways to change the values of environmental variables, such as to the constraints introduced in Step 2) that prevent IM from taking any harmful action.

Thus, we now have 29 rules, each represented in the above form. We call this the IM rule base (**IM-RB**)—the set of rules from which we want to derive policies. The technically savvy reader will note that these rules are "propositional logic" rules even though they appear to be first order rules because no variables appear in any of the TP-rules presented in this paper (Mendelson 2009).

8.2 Computing Policies

Suppose *Body* (**IM-RB**) is the set of all literals (positive and negative) that occur in the body of any rule in **IM-RB**.

Recall, as usual, that a pair of literals (L, $\sim L$) are called *complementary literals.* For instance, *kill_IM_leaders* and ~ *kill_IM_leaders* is a complementary pair, and each of these two literals is the complement of the other one.

Let *CompBody* (**IM-RB**) be the set of all literals $\{\sim L \mid L$ is in *Body* (**IM-RB**)$\}$. In other words, *CompBody* (**IM-RB**) is the complement of every literal that occurs in the body of any of the rules in **IM-RB.**

Formally, a *policy P that potentially eliminates all violent acts*[2] *of IM* (or just *policy*, for short, as it is understood that the goal of this book is to rein in IM's

---

[1] For several reasons, some steps involved in our prior study of LeT (Subrahmanian et al. 2012) were not needed in this one.

[2] Only violent acts considered in this book are included here.

terrorist actions) is a consistent subset of *CompBody* (**IM-RB**) that satisfies two conditions:

1. For each rule *r* in **IM-RB**, there is a literal in *P* whose complement also occurs in the body of rule *r* and
2. There is no *strict subset P'* of *P* satisfying the preceding two conditions.[3]

Intuitively, a policy is a set of literals denoting actions that any organization wanting to "rein in" IM might consider taking. From a technical perspective, a policy can be viewed as a "minimal model" of a logic program (Minker 1982) that cannot make the head of any rule true and that must, in addition, satisfy environmental variable constraints. Minimal model computations have been studied extensively over the years, and a variety of fast algorithms to find them have been produced (Bell et al. 1994a, b; Subrahmanian et al. 1995).

Structurally, a policy must satisfy several requirements:

1. A policy must only consist of literals in *CompBody* (**IM-RB**) . Note that each literal in *CompBody* (**IM-RB**) prevents at least one rule in **IM-RB** from firing (and possibly more than one).
2. A policy must be consistent; it cannot contain both L and $\sim L$ for any literal L.
3. A policy must prevent each rule in the **IM-RB** rule base from firing. Thus, a policy ensures that there is *no way* for IM to carry out any of the violent acts that they have carried out in the past.
4. A policy must be *minimal*. It must not recommend more actions to be taken by appropriate decision makers than are strictly necessary.

8.3 Computing Policies to Potentially Eliminate (Most) Violent Acts by IM

We use integer linear programming to solve the problem of violating all violent acts carried out by IM (that are studied in this book) with the exception of some attacks on holidays.

Let *Literals* be the set of all literals L that appear anywhere in any rule (head or body) in **IM-RB**. Let X_L be a variable (whose value is unknown) telling us if literal L is included in a policy or not. We now define a set of linear constraints *LC(IM)* as follows.

1. For each rule $A \leftarrow B_1$ & .... & B_n in **IM-RB**, *LC(IM)* contains the constraint $X_A + \sum_{i=1}^{n}(1 - X_{B_i}) \geq 1$. This constraint says that either A is true or one of the B_i's is false.

[3] When reining in IM, we did not impose constraints on what a counter-terrorism organization might do as we did in our prior study of LeT (Subrahmanian et al. 2012).

2. For each environmental variable constraint saying that the value of a particular environmental atom A is set to a fixed value c (0 *denoting false, or* 1 *denoting true*), *LC*(*IM*) contains the constraint $X_A = c$.
3. For each pair of complementary literal L, $\sim L$, *LC*(*IM*) contains the constraint $X_L + X_{\sim L} \leq 1$. This constraint says that at most one of L or $\sim L$ is true.
4. If A occurs in the head of any rule in **IM-RB**, then *LC*(*IM*) contains the constraint $X_A = 0$. This constraint says that IM cannot carry out any violent activities.
5. *LC*(*IM*) contains a constraint $X_L \in \{0, 1\}$ indicating that all variables are either 0 or 1.

The following important theorem tells us that we can find policies that potentially prevent IM from carrying out any of the violent acts for which we derived TP-rules in Chaps. 4–7.

Theorem. *LC*(*IM*) is solvable.

The following algorithm tells us how to generate *all* policies that potentially prevent IM from carrying out any of the violent acts for which we derived TP-rules in Chaps. 4–7. The Policy Computation Algorithm works as follows.

- It solves the integer linear program shown at the beginning of the while loop using any standard integer linear program solver.
- If the integer linear program is solvable, then it has found a set S of literals in the rule bodies which, if negated, would prevent every rule in **IM-RB** from firing. The set $\{\sim L \mid L \textit{ is in } S\}$ is therefore a policy. It adds this to the set of policies found thus far. Moreover, to ensure that no more policies that are supersets of this one are found by the algorithm, it adds a constraint to the set of constraints.
- If the integer linear program is unsolvable, then we have found all policies that potentially eliminate violent attacks by IM.

POLICY COMPUTATION ALGORITHM

Policies = {}; (* no policies found so far *)
Solvable = true;
Constraints = *LC*(*IM*);

While Solvable **do**
$S =$ **minimize** $\sum_{L \textit{ occurs in the body of a rule in LeTRedRB}} X_L$;
Subject to *Constraints*;

If a solution exists **then**
Policies = Policies ***U*** $\{\sim L \mid L \textit{ is in } S\}$;
Constraints = Constraints ***U*** $\{\sum_{L \in S} X_L \leq card(S)\}$
Else Solvable = false;

Return Policies

Theorem. The Policy Computation Algorithm finds all policies that potentially eliminate violent attacks by IM.

We ran the Policy Computation Algorithm and generated exactly one policy that potentially eliminates violent attacks by IM. The Chap. 9 will discuss these policies in greater detail.

8.4 Note on Alternate Policy Computation Methodologies

In this section, we discuss the intuition behind the formulation of the Policy Computation Algorithm given above and compare it to other possible, equivalent approaches. We also briefly summarize directions one could take to further improve the theoretical and empirical power of this algorithm.

We use an integer linear programming approach. This decision was motivated by both the simplicity and generality of the formulation and, importantly, the availability of efficient software packages to solve such problems. The industry standard integer programming software suites—IBM's ILOG CPLEX or the Gurobi Optimizer—are capable of solving incredibly complicated integer programs through a mix of highly specialized pre-solving, customized cut generation, and efficient tree search.

Still, those familiar with optimization (and complexity theory) in computer science will recognize that our problem is susceptible to other formulations. For example, the basic hitting set calculation is reducible to simply solving a variant of the Boolean satisfiability problem, more colloquially known as SAT. Cook (1971) and Karp (1972) provide excellent descriptions of complexity reductions. Efficient, but less industrially honed, SAT solvers exist; currently, the academic leaders include the serial MiniSAT (Eén and Sörensson 2004) or the parallel ManySAT (Hamadi et al. 2009), both of which can solve SAT formulae with millions of variables. These could have been used to produce identical results to those in this chapter and the next.

Extending the model. One advantage to our integer programming approach is the ability to easily add *weights* to the variables X_L representing whether or not a literal L is included in a policy. Intuitively, if a literal L (say, *kill_IM_leaders*) is given a large weight w_L, then that literal is "hard" for a policy maker to implement in reality. Similarly, if a literal L' (say, *no_government_ban*) has a smaller weight $w_{L'}$, then it is "easier" to effect change relative to that literal in reality. *Given* a mapping of literals to weights, we could then—through the addition of these weights to our objective function—minimize the overall *weight* of active variables rather than the overall cardinality. This would allow policymakers to use their real-world knowledge of the intricacies of diplomacy to personalize these automatically generated policy recommendations.

Scaling to larger rule bases. In this book, we explored in detail a set of 29 rules pertaining to violent attacks performed by IM. One could imagine, however, modeling a much broader situation (e.g., all terrorist groups in the region or the world), resulting in an enormous set of automatically generated rules. For these larger problem instances, the computational complexity of computing optimal policies (either by solving the integer linear program or otherwise) becomes overwhelmingly large. In these experimentally intractable cases, one could find approximate solutions—although, as we now discuss, this is a theoretically difficult problem.

The hitting set problem is, intuitively, closely related to the well-studied NP-Complete set covering problem (in effect, the problem is *equivalent* to set cover through a simple reduction using bipartite graphs). It has been shown that the set covering problem is inapproximable in theory (Lund and Yannakakis 1994). In fact, the best any approximation algorithm can do in theory is roughly equivalent to the standard greedy algorithm for approximating set cover and hitting sets, which can be off from optimal by a log factor of the number of rules considered (Alon et al. 2006). In practice, however, such approximation algorithms often lead to very good—and, importantly, tractable—approximate results. Furthermore, being able to approximately solve very large models could outweigh being limited to smaller models, even if these small models could be solved to optimality.

Dealing with infeasibility. Actors on the world stage are self-interested but rarely rational, at least in the game-theoretic sense. As we saw in our prior work on LeT (Subrahmanian et al. 2012), this lack of rationality may lead to unpredictable or seemingly contradictory actions, which, in turn, leads to infeasibility of the hitting set problem. In the event of infeasibility, a policy analyst could relax the objective function to instead automatically calculate the *largest* set of rules that could possibly be prevented from firing, rather than requiring that *all* rules do not fire. These are closely related to the concept of *maximally consistent subsets* of logic programs (Baral et al. 1992). A technique to do this using integer linear programming is suggested by Bell et al. (1994b). This problem, at least without variable weights, is also equivalent to the maximum Boolean satisfiability problem (or MAX-SAT), another well-known NP-Complete problem. Expert knowledge could also be used to cull certain rules from the larger rule base.

8.5 Conclusion

Policy makers, security analysts, and military decision makers in many countries (primarily India, but the U.S. and EU as well) have struggled with the problem of eliminating violent acts by IM. Despite much effort during the past several years, IM remains a capable organization, continuing to launch deadly attacks. One may argue, in fact, that IM has grown stronger during this time, carrying out increasingly bold and deadly violent acts across a wide variety of cities in India including Delhi, Mumbai, Hyderabad, Bengaluru, and Varanasi.

In this brief chapter, we describe our policy analytics methods that merge important elements of mathematical logic, logic programming, and integer linear programming. We apply this policy analytics methodology to computing strategies that may help rein in IM.

We then develop and describe the Policy Computation Algorithm to find all such policies, leveraging an algorithm due to Bell et al. (1994a, b). That algorithm, used to compute "minimal models" of logic programs, is adapted to compute policies.

We conclude with some caveats: the policies generated by our algorithm only *potentially* eliminate IM attacks. They do not guarantee elimination. What they do guarantee is that the policies generate an environment "around" IM that is maximally conducive to reduced attacks.

We describe the policies generated by our Policy Computation Algorithm in detail in the Chap. 9.

References

Alon N, Moshkovitz D, Safra S (2006) Algorithmic construction of sets for k-restrictions. ACM Trans. Algorithms 2: 153–177.

Baral C, Kraus S, Minker J, Subrahmanian VS (1992) Combining knowledge bases consisting of first order theories. Computational Intelligence 8:45–71.

Bell C, Nerode A, Ng R, Subrahmanian VS (1994a) Implementing deductive databases by mixed integer programming. ACM Transactions on Database Systems 21:238–269.

Bell C, Nerode A, Ng R, Subrahmanian VS (1994b) Mixed integer methods for computing non-monotonic deductive databases. J ACM 41:1178–1215.

Cook S (1971) The complexity of theorem proving procedures. In Proceedings of the third annual ACM symposium on theory of computing 151–158.

Eén N, Sörensson N (2004) An extensible SAT-solver. In Theory and Applications of Satisfiability Testing, Springer, Berlin.

Hamadi Y, Jabbour S, Sais L (2009) ManySAT: a parallel SAT solver. Int J Satisfiability, Boolean Modeling and Computation 6.

Karp R (1972) reducibility among combinatorial problems. In: Miller RE, Thatcher JW (eds) Complexity of computer computations. Plenum, New York.

Lund C, Yannakakis M (1994) On the hardness of approximating minimization problems. J ACM 41:960–981.

Mendelson E (2009) Introduction to mathematical logic, 5th Edition. Chapman and Hall/CRC Press.

Minker J (1982) On indefinite databases and the closed-world assumption. Proc 6th international conference on automated deduction, lecture notes in computer science 138:292-308.

Subrahmanian VS, Nau DS, Vago C (1995) WFS + branch and bound = stable models, IEEE transactions on knowledge and data engineering 7:362–377.

Subrahmanian VS, Mannes A, Shakarian J, Sliva A, Dickerson J (2012) Computational analysis of terrorist groups: Lashkar-e-Taiba. Springer, New York.

Chapter 9
Suggested Counter-Terrorism Policies

Abstract In this chapter, we discuss policy options towards IM. Our Policy Computation Algorithm (PCA) generated exactly one policy that has the potential to reduce terrorist attacks carried out by IM. This one policy, however, can be implemented in many different ways. This chapter presents the policy that the PCA generated, along with a set of options on how this one policy may be implemented as well as the pros and cons of these options and recommendations on the way forward.

Our data on IM include a set of environmental variables and a set of action variables. The environmental variables consist of the social, cultural, political, religious, and economic factors comprising the environment within which IM operates. Environmental variables also include actions taken by actors other than IM that shape IM's environment, such as Pakistan's ISI and the terrorist groups LeT, HUJI, and HuM.

As organizations react to changes in their environment, we also consider actions taken by India's law enforcement and counter-terrorism agencies such as the Central Bureau of Investigation (CBI), the Intelligence Bureau (IB), the Research & Analysis Wing (R&AW), the National Counter-terrorism Center (NCTC-India) of India,[1] the National Intelligence Grid (NatGrid), paramilitary forces like the Border Security Force and the Central Reserve Police Force, as well as state and local law enforcement agencies.

As described in Chap. 8, the environmental variables offer policy levers to the security establishment. By "resetting" these environmental variables appropriately, security organizations can reshape the environment in which IM operates, thus potentially inducing different (and hopefully better) behavior from them.

The variables predicting IM attacks are summarized in Table 9.1.

---

[1] The U.S. has an identically named agency.

V. S. Subrahmanian et al., *Indian Mujahideen*, Terrorism, Security, and Computation, DOI: 10.1007/978-3-319-02818-7_9,

Table 9.1 Relationships between various environmental variables and IM attacks

	Attacks on public structures	Bombings	Consecutive or timed attacks	Persons killed
Arrests of IM personnel	Yes	Yes	Yes	Yes
IM communications	Yes	Yes		Yes
IM conferences	Yes	Yes	Yes	Yes
IM claims of responsibility	Yes	Yes		Yes
Warming of relations between India/Pakistan	Yes	Yes		Yes
Internal conflict within India		Yes		Yes
IM personnel also in other NSAGs	Yes	Yes	Yes	Yes

In order to stop IM attacks, the Policy Computation Algorithm tells us that we must reshape the environment in which IM operates so as to make each of the above variables "false".[2]

The Policy Computation Algorithm, when applied to our IM dataset, only returned one policy. In some ways, this is better than results in our study of LeT (Subrahmanian et al. 2012) in which the PCA found no policies that would mitigate attacks by LeT. When attacks on holidays were excluded, however, the PCA generated eight policies that would mitigate attacks by LeT (Subrahmanian et al. 2012). In contrast, in the case of IM, we were able to find only one policy.

The one policy generated by the PCA for mitigating IM's terrorist attacks required the following actions. We note that Indian security services may already be performing many of these operations.

- *Heighted Vigilance after Arrests of IM personnel.* As in the case of a previous study of LeT (Subrahmanian et al. 2012), arrests of IM personnel are followed a few months later by every type of attack studied in this book. We do not suggest that Indian and other law enforcement and security organizations stop arresting IM operatives, but such arrests need to be followed by heightened security and intelligence gathering operations in order to disrupt the IM operations that in the past have routinely followed such arrests.
- *Monitoring IM Communication Campaigns.* With the exception of consecutive/timed attacks by IM, all other types of attacks covered are preceded by a discussion of IM's goals and capabilities, i.e., its campaign. Usually these messages are contained within the manifestos IM issues after its attacks. When these statements emphasize that IM will continue to launch attacks, they should be viewed as harbingers of plots already in the works. Indian security should monitor these statements carefully and increase security when and where appropriate.

[2] This however is not practical. For example, there are numerous reasons why better diplomatic relations between India and Pakistan would be good for both nations and for international security in general.

- *Infiltrating and Monitoring Conferences Organized by IM.* When IM organizes a conference (almost always a clandestine meeting), we can expect attacks a few months later with very high probability. This is the case for all types of attacks in the preceding chapters. It seems that operationally, IM uses these conferences to either plan attacks, select which of many possible attacks to actually carry out, or give a final green light to launch such attacks. Indian security should therefore go to great efforts to learn when such meetings occur. If such information can be obtained beforehand, conferences should either be stopped or monitored in order to disrupt attacks. If the knowledge that a conference has occurred is obtained after it has occurred, Indian security should assume attacks are in the works and increase security and attempt to disrupt plots.
- *Enhanced Vigilance after Claims of Responsibility.* When IM claims responsibility for an attack, the TP-rules show that there is an extremely high probability that all types of attack except for consecutive/timed attacks will follow within the next few months. Security organizations therefore should continuously monitor open sources and any classified datasets for any claims of responsibility issued by IM, including claims that do not appear credible. In addition, news organizations should be legally obliged to report any such claims of responsibility to law enforcement authorities.
- *Increased Vigilance after Warming of Diplomatic Relations between India and Pakistan.* Table 9.1 shows that a warming of diplomatic relations between India and Pakistan is almost always followed by all types of attacks studied (with the exception of consecutive/timed attacks). The policy does not say such diplomatic efforts should be stopped, just that when diplomatic ties between India and Pakistan are on the upswing, then intelligence gathering with regard to IM and its sponsors needs to be significantly increased and additional resources poured into disrupting potential IM attacks.
- *Increased Vigilance after Internal Conflict within India.* When there is internal religious conflict within India resulting in casualties, IM bombings are likely to follow shortly thereafter. Typically, such "internal conflict" refers to religious violence, usually between Hindus and Muslims. Law enforcement must be on a continuous watch for Hindu-Muslim violence and when such violence occurs, law enforcement must immediately step up intelligence gathering operations in order to disrupt subsequent attacks by IM. The Indian security establishment is well aware of the relationship between religious violence in India and subsequent attacks.
- *Monitoring Membership of IM operatives in other non-state armed groups (NSAGs).* When IM operatives are shared with other NSAGs they are involved in each and every type of attack studied in this book.[3] As discussed in previous chapters, these links are essential for IM to gain the knowledge and skills

[3] We note that membership of IM operatives in other NSAGs is a near-constant condition. We remind the reader that this variable was coded as being true when news reports explicitly noted shared membership between IM and other groups.

Table 9.2 Proposed strategic actions

Strategic policy action	Tactical implementation
Reduce likelihood of attacks after arrests of IM personnel	Improve acquisition of intelligence obtained from suspects and better aggregation of this information with other intelligence on IM activities
	Increase security at likely targets after IM arrests
	Reduce publicity given to arrests of IM personnel
Monitor IM communications for discussions of its terror campaign	Increase security and intelligence activities when IM's manifestos discuss future operations
	Develop assets and methods to counteract IM messaging intended to energize its cadres
Monitor conferences organized by IM	Increase infiltration into IM networks, intelligence gathering, and electronic surveillance of IM activity
	Build a "travel intelligence" capacity to systematically monitor the movements of IM members to learn when meetings are occurring and if possible disrupt them
Monitor claims of responsibility by IM	Monitor news and social media for claims of responsibility and develop tools to identify origins of claims in real-time
	Enact legislative tools needed to require news organizations to report claims of responsibility and that state, local, and national security forces share this information in a timely manner
Internal conflict within India	Develop tools to monitor Hindu-Muslim sentiment across multiple data channels including news, blogs, social media, SMS, and phone calls
	Develop security capabilities to defuse potential communal violence
Warming of diplomatic relations between India and Pakistan	Increased infiltration of IM network and increased intelligence gathering and electronic surveillance of IM activity
	Develop tools to continuously monitor sentiment in news, blogs, social media on India–Pakistan relations, Hindu-Muslim relations, and leadership of India and Pakistan

necessary to carry out attacks. Intelligence operations are needed to disrupt the ties that enable other NSAGs to help IM. Specifically, intelligence needs to be continuously gathered on the links between IM and others NSAGs and tactics need to be developed and deployed to sever these links.

Table 9.2 summarizes the proposed strategic actions and how they address the variables correlated with IM attacks of different types. We also suggest tactical methods to implement these policy options. Again, we note that Indian security agencies are probably doing much already towards these goals.

The rest of this chapter examines several options for implementing each of these options on the ground.

9.1 Arrests of IM Personnel

When IM personnel are arrested by Indian security organizations, there is a high probability of terrorist strikes within the next few months. The strikes include bombings, attacks on public sites (e.g. markets), and timed/consecutive attacks.

We are not suggesting that law enforcement and security organizations stop arresting IM personnel. There are many good tactical reasons for why this could be important. For instance, Indian law enforcement may receive intelligence that certain IM operatives are gathering in a certain place—arresting them and extracting intelligence from these IM operatives may be crucial to identifying planned operations and disrupting these attacks.

While arrests *may* lead to a long-term deterioration of IM capabilities, our model indicates that *arrests of IM operatives do not disrupt IM from carrying out further attacks* within a few months after those arrests. In fact, arrests are almost always followed by the different types of attacks. One possible explanation for this counter-intuitive finding is that these arrests inspire IM cadres. Another is that with the arrests, IM operatives grow concerned that those arrested will reveal their plans and accelerate operations. Another possibility is that the increased level of action and communication involved in planning terror attacks is flagged by the Indian security establishment who carry out arrests of some of those responsible.

Indian law enforcement often publishes details of arrests of IM personnel. This is natural—most law enforcement agencies in the world, including those of the U.S., want to publicize their successes. There is also a public morale component to publicizing the details of the arrests of terrorists. In the wake of terror attacks, the public needs to be reassured that security services are capable of preserving order.

In Israel, debates over counter-terror policies incorporated this issue and, at times, the need to boost public morale was an important factor in making decisions about counter-terror policy. But the Israelis ultimately determined that boosting morale, in and of itself, was not a sufficient reason to carry out a counter-terror action (Ganor 2005). This logic applies to India in two regards. The first consideration is whether or not to carry out a campaign of arrests at all. Some of the arrests may be large-scale round-ups of people with limited connections to suspects and in other cases intelligence gathering might be better served by monitoring suspected individuals. The second aspect is whether or not to publicize the arrests. These arrests may boost public morale but may also contribute to future attacks as IM reacts violently in order to energize their own cadres. Additionally, the arrests might reduce Indian security's ability to prevent future attacks by not allowing the threat of arrest to "turn" certain IM operatives into informers.

Recommendation—Arrests with minimal publicity. Thus, a first recommendation is that when IM operatives are arrested, this be done "away" from the limelight and outside the public view. Of course, the law in most countries including India requires that due process be followed when individuals are arrested. Special counter-terrorism courts should be instituted where such arrested IM operatives are given due process under India's constitution out of the glare of India's energetic press. Further procedures should be established and promulgated throughout India's law enforcement community on media relations to ensure the public is informed without undermining ongoing operations.

Recommendation—accompany arrests with misinformation campaigns. A second recommendation is that security organizations should put out deliberate misinformation (to the extent allowed by law) when IM operatives are arrested so that IM itself is unsure of what happened to the arrested individuals. In other words, when an IM operative is arrested, information operations should be carried out that leaves IM guessing about the true status of the individual. Was he arrested? Was he really working for IM or was he an intelligence operative working for Indian security? Or did he just want to terminate his relationship with IM? With appropriate intelligence operations and under appropriate legal safeguards, security officials can compromise the arrested individual's emails, phone records, personal correspondence, travel schedules, and even financial records to build a credible "cover story" for why the person disappeared. Such operations have the potential to keep IM off-balance, wondering how much their own planning and operations have been penetrated or compromised, thus forcing them to expend resources that they would have otherwise spent on planning terrorist attacks.

Recommendation—surveillance and intelligence gathering operations after arrests should be stepped up. After arrests or covered up arrests as suggested above, IM intelligence gathering and surveillance operations should be significantly increased. Indian security services should develop stronger capabilities for exploiting intelligence from recent arrests and incorporating that data into their broader picture of IM operations.

Recommendation—detailed post-arrest intelligence assessments and integrated security data systems should be done. After arrests of IM personnel, a detailed analysis of relevant intelligence of IM should be done to identify likely targets of subsequent IM attacks. Methods to randomly assign enhanced security patrols to such places should be significantly increased. Computational tools for randomly deploying police and law enforcement units around a set of possible targets can be used to keep attackers unsure about whether they can successfully attack certain targets (Dickerson 2010).

A major problem however in implementing this recommendation is that India's law enforcement and security agencies are dispersed and not well integrated. For instance, state law enforcement agencies often do not share information with national law enforcement and security agencies. Likewise, even within the central government, data about an individual or a network can take time to filter through to other organizations. A single national real-time seamless information system that crosses the standard organizational boundaries is needed so that investigators can have authorized information on a real-time basis (Gordon 2010). India has already recognized the need for such a system. The Union Home Ministry in Delhi has proposed a framework called the Crime and Criminal Tracking Network and Systems (CCTNS), but this does not appear to be fully operational at this time.[4] The operational deployment of CCTNS should be an urgent national priority.

Recommendation—Counter-terror capabilities should be improved. This recommendation focuses on training. The 26/11 Mumbai attacks highlighted the lack of trained personnel and the inability to deploy resources to the theater of operations in a timely manner. India's state of preparedness has improved substantially since the Mumbai attacks, but there is no such thing as being "too prepared". Indian security forces face many challenges, from local law enforcement to national para-military services (Fair 2012; Gordon 2010). This recommendation really applies to all the policy options above. Reliable predictions of terror attacks from computational tools do little good if the government lacks sufficient tools to increase security and disrupt attacks. We recommend therefore that joint counter-terrorism training operations be instituted in which Indian counter-terrorism forces work closely with the U.S., Israel and other allies to build counter-terror capabilities.[5] Their expertise and experience will be invaluable to Indian counter-terrorism forces charged with nullifying threats posed by IM and other organizations. The U.S. and other allies will also benefit from India's experience and gain a better understanding of the operational environment of India's security agencies, ultimately forging stronger bonds for future counter-terror cooperation.

9.2 IM Communication About Terror Campaign

When IM communiqués discuss their terrorism campaigns, there is a high probability of carrying out terrorist strikes within the next few months. As an underground organization, IM does not carry out overt communications campaigns but rather includes manifestos of its grievances and of its capabilities and intentions for further attacks in its claims of responsibility (Swami 2008a, b). These

[4] As this book is going to press the program's website (http://ncrb.nic.in/cctns.htm) indicates that the CCTNS is still in the process of being implemented.

[5] An overview of U.S. counter-terror capacity building programs can be found in Kraft (2012).

manifestos effectively state that more violence is coming. These manifestos are a source of intelligence in and of themselves and should be analyzed systematically.

Recommendation—Communications from IM must be subject to immediate real-time analysis for signs of future attacks, and alerts should be issued accordingly. Both human analysts and computational tools should carefully scrutinize IM communiqués and other communications material. As discussed above, when IM implies that there will be future attacks they often deliver on this promise. Given these warnings, Indian security forces should be alert for developments indicating a possible plot. Improved intelligence capabilities and security measures would help Indian security forces prepare in the event of these warnings.

9.3 IM Conferences

When IM holds a conference (nearly always in secret), there is a very high probability that the conference will be followed a few months later by violent attacks. Table 9.1 shows that a few months after holding a conference, IM launches all kinds of attack discussed in this book within a period of 1–5 months. These meetings are essential for planning and training operations; without these meetings, terrorists would be hampered in preparing and coordinating complex operations (Forest 2006; 9/11 Commission 2004).

Preventing IM leaders from meeting or monitoring these meetings should be a critical goal of Indian domestic security forces. Doing so is difficult because these meetings are held clandestinely. Nonetheless, India should invest in developing the capabilities needed to determine when these secret meetings are occurring. Ideally, terrorists should be prevented from meeting at all, but if that is not possible, detecting and monitoring these meetings (and if possible infiltrating them) would yield critical intelligence, and even determining that a meeting occurred after the fact would at least be an indicator that a future attack is likely and give security forces some warning and opportunity to prepare.

Recommendation—Establish travel intelligence capabilities. Indian security forces need to develop a range of sophisticated intelligence capabilities to identify and disrupt terrorist meetings. One critical capability, identified by the 9/11 Commission, was that of travel intelligence, so that security agencies can counter the ability of terrorists to move, communicate, and meet without being detected. India also needs to improve its ability to gather and blend intelligence from various sources and disseminate relevant warnings in real-time.

Recommendation—Disrupt IM operations in the wake of conferences. Knowledge that a terrorist conference had occurred creates several possible options for disruptive operations:

- Removal of key IM personnel from the theater of operations. This could mean detention as allowed under appropriate laws of the jurisdictions involved or elimination if authorized by a court. It could even mean the provision of amnesty, safe passage and/or financial incentives, in exchange for providing detailed intelligence on planned IM operations and other details. Spezzano (2013) discusses methods to identify which players within a terrorist network should be removed in order to minimize the expected lethality of the resulting network;

- Cyber-operations to leave other IM cadres unsure about the state of the removed individual. Ideally, cyber-operations would leave remaining IM members to wonder if he left or if he provided intelligence on IM plans and operations. Such cyber activities could include leaving compromising correspondence, ambiguous email, or SMS trails that IM is likely to find, as well as financial transactions that may compromise the individual in question. U.S. intelligence supposedly carried out such operations against the Fuerzas Armadas Revolucionarias de Colombia (FARC) in Colombia significantly disrupting their operations (Reyes and Dudley 2006).

In short, there is a need for pre- and post-conference actions, as well as methods to gather intelligence during the period of IM conferences. This book primarily focuses on strategic rather than tactical issues, but counter-terrorism officers can develop a range of tactics based on past best practices to implement these strategy options.

9.4 IM Claims of Responsibility

When IM claims responsibility for attacks, it is clear that they are emboldened. A few months after such claims of responsibility, IM launches every kind of attack discussed in this book with the exception of timed/consecutive attacks.

Perhaps IM claims of responsibility for past attacks are used as a way of energizing IM cadres. Perhaps they are used as a way of gaining legitimacy for the organization. Whatever the reason, it is clear that when IM claims responsibility for attacks, follow-up attacks are planned.

Recommendation—Systematically monitor and track IM claims of responsibility. Relevant security organizations should monitor both open-source and classified information feeds for evidence of IM claims of responsibility.

Recommendation—Pass legislation so that messages from IM to the media are reported to law enforcement in a timely manner. IM has historically claimed responsibility for attacks by sending emails and other messages to a variety of organizations, primarily the press. For example, the IM announced its existence in an emailed manifesto sent five minutes *before* blasts hit courthouses in Varanasi, Lucknow, and Faizabad. Similar manifestos have been sent to the press just before and just after other IM attacks (Gupta 2011; Ranjappa 2008). We therefore recommend that India pass appropriate legislation that requires any organization receiving such claims to pass the information to relevant law enforcement and security agencies as soon as possible.

Recommendation—Develop intelligence capabilities to quickly analyze claims of responsibility. Indian security should develop intelligence assets that can exploit any intelligence gleaned from these manifestos, including cyber-forensics (PTI 2010).

Recommendation—Develop intelligence capabilities to disrupt the IM network. Relevant intelligence agencies need to develop the assets needed to disrupt any planned IM attacks. A mix of kinetic and cyber strategies for such disruption might include methods to:

- Remove relevant individuals from the theater of operations, one way or the other, perhaps using the methods suggested in Sect. 9.3;
- Conduct covert operations to confuse IM cadres about the status of these "removed" individuals by spreading multiple messages through the IM network, e.g., the removed individuals were working for foreign intelligence agencies. This would be accomplished by spreading misinformation about their email/phone communications, by misrepresenting their financial transactions, and/or by misrepresenting their transactions with other IM personnel.

Security officials in India are almost certainly aware of the recommendations made here. Our findings merely reinforce the need to implement these well-known counter-terror tactics and strategies.

9.5 Warming of India–Pakistan Diplomatic Relations

Table 9.1 indicates that when there is a warming of diplomatic relations between India and Pakistan, there is a high probability of IM-backed attacks on Indian targets. In fact, when such diplomatic warming occurs, IM carries out attacks on public sites and bombings within a few months.

The Government of India is fully aware that forces within Pakistan seek to disrupt efforts to normalize relations with Pakistan. Table 9.1 also shows that when diplomatic relations between India and Pakistan are looking up, there is a high probability of IM attacks as IM, and their Pakistani sponsors (LeT and elements within the Pakistani military) seek to disrupt warming relations.

Thus, anytime there is an improvement in Indo-Pakistani diplomatic relationships, there is a critical need to:

- Use real-time text analytics systems such as T-REX (Albanese 2007) and Automatic Coding Engine (ACE , Albanese 2013) to ramp up surveillance of IM activity in both open-source and classified datasets[6];
- Use intelligence and surveillance assets to significantly ramp up intelligence data collection on the IM organization;

[6] A commercial system for real-time text analytics is the Listening Platform deployed by Tata Consultancy Services for numerous companies worldwide.

- Use information operations to disrupt detected IM operations to plant doubts among IM operatives about fellow conspirators using faked emails, faked financial transactions, and other faked assets to destroy the credibility of IM operatives; and
- Use network disruption techniques (Spezzano 2013) to identify key individuals to remove or turn within IM's terror network.

In addition, recent results (Dickerson 2013) and prior research (Mannes et al. 2010; Subrahmanian et al. 2010) suggest that U.S. aid to Pakistan has had no success in reducing attacks by the LeT. Elements of IM are linked to LeT (which opposes improved Indian-Pakistani relations) and its sponsors in the ISI. While some analysts see IM as strictly an arm of Pakistan's intelligence agency, others also see it as an indigenous movement that has received substantial aid from Pakistani terrorists. Regardless, Pakistan and particular sub-national actors within Pakistan cannot be ignored as a factor in understanding IM activities (Fair 2010).

Recommendation—Develop flexible sanctions and aid programs for Pakistan. We recommend that the U.S. continue to fine-tune its aid and relations with Pakistan. Often U.S. aid and sanctions are blunt instruments that inadvertently harm both the U.S. and its allies within Pakistan while aiding U.S. enemies (Mannes et al. 2010; Subrahmanian et al. 2011). Subrahmanian et al. (2012) suggest reducing U.S. aid to Pakistan in order to combat LeT operations, while supporting private NGOs to provide basic services such as health and sanitation to impoverished populations in Pakistan. Fair (2012) proposes developing more effective sanctions programs that will harm the Pakistani military, which has not cut its ties with terrorists, while aiding Pakistan's civilian government, which has pursued better ties with India. At times, U.S. aid to Pakistan has not prioritized countering extremism within Pakistan and even, inadvertently, enabled Pakistani extremists.

9.6 Monitoring Internal Religious Conflict Within India

Internal religious conflict within India is largely outside the control of the Indian government. Spontaneous attacks, such as the 2002 attacks by Gujarati Muslims on a train carrying Hindus who were traveling to show their support for building a Hindu temple at a location that had been previously the site of mosque destroyed by Hindus a decade before, occur with little warning. The attack on the train took 58 lives and sparked Hindu mobs to attack Muslim homes and businesses, killing at least a thousand. This attack, while the bloodiest example of communal violence in recent years, is not unique (Varshney 2004).

Predicting and pre-empting spontaneous violence is extremely difficult and requires a range of skills and capabilities. But, inter-communal violence in India appears to motivate terrorists, so developing policies and capabilities to reduce it should be a top priority for the Indian government. Besides improving training for the security forces, there are important computational capabilities that can help the Indian government identify flashpoints so that violence can potentially be forestalled.

Recommendation—Utilize real time text and sentiment analytics platform to analyze communal publications, news articles, social media posts, and tweets. Our first recommendation to law enforcement agencies is to utilize a real-time text analytics platform that scours both open-source data (e.g., news feeds, blog feeds, social media feeds, and more) and classified security feeds (e.g., intelligence cables, email and phone intercepts). Such a real-time text analytics platform would maintain comprehensive situational awareness of communal tensions and potential flashpoints. There are numerous such text analytics platforms available worldwide. Within India, Tata Consultancy Services' "Listening" platform is used by numerous customers worldwide; this and other similar platforms can be customized and used to good effect. In addition, we recommend the use of sentiment analysis platforms to identify the sentiments and emotions in such open source or classified feeds, in so far as written communications expressing certain sentiments and emotions are likely to be correlated with outbreaks of inter-communal violence. Sentiment analysis programs are also well developed and have been shown to be correlated with several phenomena such as riots and civil violence. Examples of sentiment analysis systems include OASYS (Cesarano 2007) and AVA (Reforgiato 2008; Subrahmanian 2009). Such systems can also be used to monitor chatter within communities under appropriate legal authority.

These tools can provide an early warning system by tracking significant levels of either anger towards Hindus or anger towards Muslims and can allow India's state, local, and national government to deploy security forces or take other measures to defuse the crisis.

In addition, once such sentiments and/or attack information are detected, we recommend that relevant security agencies automatically "ramp up" their level of preparedness and use all available intelligence assets to collect further data on planned IM attacks. This requires the development of improved capabilities to monitor IM and to merge different streams of intelligence and communications capabilities so these assets can provide real-time information to security agencies.

Recommendation—Develop specialized fast reaction teams to forestall communal violence. Our second recommendation is that the Government of India develop security units (and the appropriate legal and logistical infrastructure to support them) to intervene when any kind of ethnic or religious conflict erupts. IM's own manifestos identify communal violence, particularly revenge for atrocities against Muslims, as the central motivator in their attacks. Ultimately, improved local policing will help ameliorate these dangers, but it will take a substantial time to implement any reforms (Gordon 2010). In theory the central government could develop fast reaction forces for this problem in the nearer term.

Recommendation—Increase security when communal violence flares. Our third recommendation is that once a threat is detected, relevant security agencies ramp up their effective operational level so that potential threats can be thwarted through the deployment of additional security assets at potential targets. The trick here is to "throw" the attackers off the target by randomizing the protection of potential targets so that IM is never sure what targets will be protected when they launch an attack. This confounds IM planning and operations, making them waste scarce resources in an attack whose success they are not sure of. The number of potential targets in India is enormous so there are limits to this strategy. But there are some patterns to IM target selection that suggest at least some probable locations that could be better protected. Further, such exercises could be useful in improving the quality of Indian local security services which, in turn, would strengthen their ability to prevent or respond to an attack.

9.7 Disrupting IM's Relationship with Other NSAGs

All types of attack carried out by IM are linked to IM's relationship with other NSAGs. When IM shares resources with other NSAGs, they appear to be emboldened. Perhaps other NSAGs share expertise and assets with IM or IM cadres receive training and logistical support from other NSAGs. For example, there is evidence that the 2006 Mumbai train bombings were carried out by IM with support from both the Pakistani ISI and LeT (Reidel 2013). Two of the IM's key founders, Riyaz Bhatkal and Sadiq Sheikh, received money and training from LeT. They then recruited more Indian Muslims, (with help from LeT and possibly Pakistani intelligence). The recruits in turn traveled from India to Pakistan via Bangladesh and the Persian Gulf to train with LeT.

Interdicting IM's operations necessarily means tracking the entire milieu of jihadist groups supported by Pakistan's ISI. Historically, IM has had close links with LeT, HuM, and HuJI, amongst others. Reliable reports (Reidel 2013) also indicate that the ISI supports other armed groups such as Lashkar-e-Jhangvi (LeJ) and Sipah-e-Sahaba within Pakistan, and the Afghan Taliban. In short, the ISI supports a number of groups and acts as a broker in enabling an armed group to secure logistical, financial, and other resources from other armed groups that ISI funds and directs.

Recommendation—Build travel intelligence capabilities and improve counter-terror intelligence sharing. In order better disrupt such transnational terrorist links, India needs to develop better intelligence capabilities to monitor terrorist travel and communications. Some of these efforts need to be internal, such as improving intelligence sharing between all levels and branches of government while others need to focus on improved international intelligence cooperation. The arrest of Abu Jundal in Saudi Arabia and his extradition to India is one important step in that direction (Tankel 2012).

Recommendation—Develop smarter and more flexible U.S. policies towards and incentives for Pakistan. U.S. policy towards Pakistan needs to be better calibrated against Pakistani support for terrorism. Sanctions for terrorist support and rewards for improved behavior need to be clearer and better targeted towards the elements of the Pakistani government relevant to the activity (Fair 2012; Subrahmanian 2010).

9.8 Conclusions

This chapter proposes a significant set of policy options to reduce the occurrence, frequency, and lethality of IM terrorist attacks. Policymakers in both India and the U.S. need to recognize that these attacks are inspired, and in many cases explicitly supported, by the ISI.

Specifically, we propose the following policy recommendations.

- *Increased electronic intelligence of IM activity.* Multiple intelligence agencies are making efforts to monitor IM communications, whether voice (phone) or text (SMS and email). Unfortunately, the gathering of electronic intelligence is no longer a national issue. IM operatives communicate with handlers in other nations, often using open communications platforms such as the Blackberry, Gmail, Yahoo mail, Skype, and other Voice Over IP systems. These communications cross national boundaries, but getting timely responses from private corporations in response to intelligence requests can be a challenge. This leads to our next recommendation.
- *Improved international counter-terrorism electronic surveillance cooperation.* There is an urgent need for a counter-terrorism framework governing electronic communications. The U.S. and India have an asymmetric relationship—major communications platforms such as Skype and Hotmail (controlled by Microsoft), Gmail (controlled by Google), Yahoo mail (controlled by Yahoo) are U.S. corporations that are responsive to U.S. law enforcement requests. In discussions with the authors, Indian government representatives have asserted that U.S. communications and Internet companies do not respond in a timely manner for counter-terrorism requests from the Indian government. Better information-sharing arrangements, which may be informal frameworks or may require international treaties with and new legislation in the US, Canada, and the EU, are needed to ensure that counter-terrorism support is provided in a real-time manner to appropriate law enforcement agencies of participating countries so as to avert terrorist attacks.
- *Reduction of publicity associated with arrests of IM personnel.* Indian security agencies should reduce the publicity of arrests of IM personnel, which are linked with subsequent attacks by IM. This may require establishing a national protocol regarding public relations that encompasses India's innumerable state, local, and national law enforcement agencies.
- *Use of text analytics and monitoring of news and social media.* U.S. and Indian security agencies must closely monitor email, news and social media in real time so as to be cognizant of threats posed by IM and the conditions that spark their attacks. When there are Hindu-Muslim conflicts within India that involve casualties, we can expect IM-backed attacks a few months later. There is a critical need to carry out ubiquitous 24/7 surveillance of online activity in India under appropriate privacy safeguards and under a transparent legislative surveillance policy. Technology to carry out "online surveillance" as well as mobile phone surveillance already exists in a number of Indian IT firms such as

Tata Consulting Services whose "Listening Platform" provides these services to a wide variety of commercial customers. In addition to analysis of the events reported in such textual objects, sentiment analysis of these text documents on a wide variety of topics must also be carried out using techniques such as those developed by Sentimetrix, Inc.

- *Use of key player analytics.* Social media, news, blogs, tweets, and SMSs must also be analyzed so as to identify the key players involved in spreading a message to IM cadres and in bridging operations between IM operatives and their associates. Once such key players are identified, they can be monitored closely using more traditional HUMINT and ELINT methods. Moreover, methods can be developed to counter-act and confound the messaging put out by pro-IM operatives, thus causing IM and its allied groups to expend resources in separating fact from fiction.
- *Enactment of legislation to require prompt reporting of claims of responsibility.* As shown in previous chapters, IM often carries out attacks a few months after claiming responsibility for an attack. This may mean that attacks are carried out in order to energize their cadres. Perhaps these attacks are meant to disrupt the warming of diplomatic relations between India and Pakistan. Without interviewing IM operatives, it is impossible to distinguish which of these motives led to IM claiming responsibility. Press organizations, the most frequent recipients of IM claims of responsibility, must be required to report these claims to law enforcement in a timely manner.
- *Creation of a central counter-terrorism data authority in India.* India is an enormous country with a huge population that is vastly more diverse in language, religion, ethnicity, and history than the U.S. and most nations in the EU. In order to preserve the rich and unique cultures within India, state governments have far greater powers than those possessed by states and provinces in other democracies. One of these powers is the enormous discretion of local police in handling security matters. This is not unreasonable, as states must be policed on the basis of local norms and cultures. But terrorist groups such as IM and LeT can take advantage of this fragmentation in law enforcement and security. We therefore recommend the creation of a *national* counter-terrorism *data* authority in India, which would centralize data about criminals and terrorists in much the same way as the U.S.'s National Counter-Terrorism Center. When serving as Home Minister, one of the Congress Party leaders, P. Chidambaram, moved forward on the creation of three entities—the Indian National Counter-Terrorism Center (we refer to this as NCTC-India to distinguish it from the US's NCTC), the National Intelligence Grid (NatGrid), and the CCTNS (Crime and Criminal Tracking Network and Systems) which centralizes data on all criminal activity in India into a central database. Making CCTNS fully operational should be a major priority for India. Yasin Bhatkal of IM was arrested in Kolkata and released when he claimed to be someone else (Indian Express 2013). Data centralization would reduce the likelihood of such missed opportunities in the future.
- *Strengthening of Covert Action.* As mentioned earlier, increased covert action is needed to distract IM's operational planners. This can take many forms. Covert

action might "remove" IM operatives from the theater of operations in India. In the past, such "removals" only included arrests and kills. Carefully designed "removal" campaigns can help significantly paralyze the IM leadership—and techniques with regards to whom to target using network analytics are now emerging (Spezzano 2013). For instance, anti-IM security agencies can attack IM by sowing deliberate misinformation when an IM member disappears. Such disinformation can include faked emails and SMSs as well as fake financial transactions intended to confuse IM about which of their operations have been compromised.

- *Increased Cooperation between U.S., Indian, and Israeli Intelligence.* The U.S., India, and Israel all face a common terrorist threat emanating from Pakistan. Destabilization of India through Pakistani-backed IM attacks on the Indian heartland could potentially be devastating to the Indian economy, which, in turn, could be devastating to the world economy. While India has lost more lives to terrorist attacks than any other country in the last 15 years (U.S. Department of State 2012), the U.S. and Israel have also been severely victimized. All of these countries have a range of expertise and experience and would benefit from improved counter-terror cooperation. There are many aspects to counter-terror cooperation, including expanded intelligence sharing, improved law enforcement and security training, forensic techniques, and technological capabilities. Stronger cooperation between India, the U.S., and Israel would be a win–win for all countries involved in the counter-terror and homeland security fields. Historical events may impose some constraints such counter-terrorism cooperation, but initial steps can be a building block to improved relations.

IM is a small, adaptable, networked organization.[7] It deploys a distributed, non-hierarchical structure, is adaptive and creative in its use of technology and has proven itself to be cyber-savvy. India needs to develop the capacity to effectively counter such organizations. While IM is perhaps India's first adversary of this type, trends indicate that it is unlikely to be the last.

References

Albanese M, Subrahmanian VS (2007) T-REX: A System for Automated Cultural Information Extraction, Proc. 2007 International Conference on Computational Cultural Dynamics, College Park, MD, pages 2–8, AAAI Press, Aug. 2007

Albanese M, Fayzullin M, Shakarian J, Subrahmanian VS (2013) Automated Coding of Decision Support Variables, In: Handbook on Computational Approaches to Counterterrorism, pages 69–80, Springer.

Arquilla J, Ronfeldt D (2001) Networks and netwars: the future of terror, crime, and militancy. RAND Corporation.

---

[7] Of the sort discussed in the modern classic *Networks and Netwars* (Arquilla and Ronfeldt 2001).

Cesarano C et al. (2007) OASYS 2: an opinion analysis system., Proc 2007 Intl Conf on Web & Social Media 313-314.

Dickerson J, Simari G, Kraus S, Subrahmanian VS (2010) A graph-theoretic approach to protecting static and moving targets from adversaries, Proc 2010 Intl Conf on Autonomous Agents and Multi-Agent Systems: 299-306.

Dickerson J, Sawant A, Hajiaghayi MT, Subrahmanian VS (2013) PREVE: A Policy Recommendation Engine based on Vector Equilibria applied to reducing LeT's attacks, Proc. 2013 International conference on foundations of open source Intelligence–security informatics (FOSINT-SI 2013), Niagara Falls, Canada

Fair CC (2010) Students Islamic Movement of India and the Indian Mujahideen: an assessment. Asia Policy 9 January.

Fair CC (2012) What to do about Pakistan. Foreign Policy June 21 http://www.foreignpolicy.com/articles/2012/06/21/what_to_do_about_pakistan.

Forest JJ (2006) Teaching terror: strategy and tactical learning in the terrorist world, Rowman and Littlefield

Ganor B (2005) The counter-terrorism puzzle: a guide for decision makers. Transaction Publishers.

Gordon S (2010) India's unfinished security revolution. Institute for Defence Studies and Analysis 11 http://www.idsa.in/system/files/OP10_IndiasUnfinishedSecurityRevolution_0.pdf.

Gupta S (2011) Indian Mujahideen: the enemy within. Hachette, Gurgaon.

Indian Express (2013) Yasin Bhatkal walked out of Kolkata jail as Bulla Mallik. Indian Express February 24. http://www.indianexpress.com/news/yasin-bhatkal-walked-out-of-kolkata-jail-as-bulla-mallik/1078815/.

Kraft M, Realuyo C (2012) U.S. interagency efforts to combat international terrorism through foreign capacity building programs. In: Weitz R (ed) Project for National Security Reform Case Studies Working Group Report 2 Strategic Studies Institute.

Mannes A, Silva R, Subrahmanian VS (2010) More military aid to Pakistan? The Huffington Post, November 22.

Nanjappa V (2008) Minutes before blasts email said: 'Stop us if you can.' Rediff.com July 26. http://www.rediff.com/news/2008/jul/26ahd3.htm.

Press Trust of India (2010) Indian Mujahideen had hacked internet connection of Navi Mumbai resident. NDAIndia.com December 8. http://www.dnaindia.com/india/report_indian-mujahideen-had-hacked-internet-connection-of-navi-mumbai-resident_1478244.

Reforgiato D, Subrahmanian VS (2008) AVA: adjective verb adverb combinations for sentiment analysis. IEEE Intelligent Systems 23:43-50.

Reidel B (2013) Avoiding Armageddon: America, India, and Pakistan to the brink and back. Brookings, Washington DC.

Reyes G, Dudley S (2006) "Ex-con helps U.S. deliver satellite phones to FARC," Miami Herald, May 10

Roul A (2009) India's home-grown jihadi threat: a profile of the Indian Mujahideen, Jamestown Foundation Terrorism Monitor March 3. http://www.jamestown.org/single/?no_cache=1&tx_ttnews[tt_news]=34577&tx_ttnews[backPid]=7&cHash=7d49c63104.

Spezzano F, Subrahmanian VS, Mannes A. (2013) STONE: Shaping Terrorist Organizational Network Efficiency, Proc. 2013 IEEE/ACM International Conference on Advances in Social Network Analysis and Mining, Niagara Falls, Canada, August 2013.

Subrahmanian VS (2009) Mining online opinions. IEEE Computer July.

Subrahmanian VS, Mannes A. (2010) Black hole for foreign aid. Washington Times September 24.

Subrahmanian VS, Mannes A, Shakarian J, Sliva A, Dickerson J (2012) Computational analysis of terrorist groups: Lashkar-e-Taiba. Springer, New York.

Swami P (2008) Not just a claim, a manifesto for jihad. The Hindu May 17. http://www.hindu.com/2008/05/17/stories/2008051754761100.htm.

Swami P (2008) Pakistan and the Lashkar's jihad in India. The Hindu December 9. http://www.hindu.com/2008/12/09/stories/2008120955670800.htm.

Tankel S (2012) The Mumbai blame game. Foreign Policy. July 9. http://afpak.foreignpolicy.com/posts/2012/07/09/the_blame_game.

The 9/11 Commission (2004) The final report of the national commission on terrorist attacks 265 upon the United States. July 22. http://www.9-11commission.gov/

U.S. Department of State (2012) Country reports on terrorism 2011. http://www.state.gov/documents/organization/195768.pdf.

Varshney A (2004) Understanding Gujarat violence. Contemporary conflicts. http://conconflicts.ssrc.org/archives/gujarat/varshney/index.html#e1.

Chapter 10
Building a National Counter-Terrorism Center

Abstract A major operational problem in countering terror attacks in India is the lack of a coordinated authority or even a central database of all arrested individuals, suspects, or ongoing operations. A second problem is the lack of a single operational entity responsible for fighting terrorism. Because information about counter-terror operations is spread across multiple states and organizations, most security organizations in India do not have all the information they need in order to fight such a war. This chapter focuses on the history, challenges, and future of a national counter-terrorism center in India, similar to the U.S. NCTC created in the aftermath of 9/11.

Every act of terrorism is a challenge to constitutional authority. It is a failure of the national government to protect its citizens from hostile elements that aim to disrupt public peace and demoralize the community. Terror succeeds when law enforcement agencies either lack the necessary advance intelligence on the conspiracy or fail to appreciate the data already available. At first sight, such data points appear unrelated, but when collated with patience and skill, they can illuminate the bigger picture of a terrorist plan. "Intelligence failure" is a refrain often heard after terror strikes. Modern history is replete with instances of these two types of intelligence failure:

- lack of advance information.
- failure to see the whole picture from available facts.

The September 11 attack is an example of how at least two major U.S. intelligence agencies neglected to share information already collected and how these agencies failed to "connect the dots." Had the collected information been shared, U.S. agencies might have suspected the rationale behind a few odd individuals with no connection to aviation enrolling in flight instruction courses at separate centers. If this anomalous conduct had been flagged, an alert might have been issued that could have unraveled the 9/11 conspiracy.

V. S. Subrahmanian et al., *Indian Mujahideen*, Terrorism, Security, and Computation, DOI: 10.1007/978-3-319-02818-7_10,

In the ISI-backed attack on Mumbai (November 26, 2008) (hereinafter referred to as 26/11), a terrorist associate in Chicago, David Coleman Headley (aka Daood Sayed Gilani), had similarly come to police notice through information from his wife. But no action was taken against him. Headley continued to aid the groups planning the attack but his visit to Mumbai was not flagged by the Indian State or central intelligence agencies. Similarly, in the February 21, 2013 explosions in the commercial area of Dilkushnagar of Hyderabad (India), a preliminary survey of the targeted area by one of the conspirators had gone unnoticed.

Along the same lines, news sources in India report that IM's operational commander in India, Yasin Bhatkal, was arrested for theft by Kolkata police in December 2009. At that time, he was already wanted for the August 2007 Hyderabad blasts as well as the September 2008 Delhi blasts. But Bhatkal reportedly convinced Kolkata police that he was an innocent civilian called Bulla Mallik and was released after a few days (Indian Express 2013). This situation periodically occurs in different parts of the world when police are unable to confirm that a person in their custody is wanted for a separate crime than the one for which he is being held.[1] In most of these cases, however, the criminal is wanted for a petty crime. In contrast, Yasin Bhatkal was one of India's most wanted criminals. The fault lay not with Kolkata police (despite criticism being leveled at them for it), but rather with a fragmented Indian security establishment with poor information sharing across national, state, local, and organizational boundaries.

In these cases, and in many other terror attacks across the globe, intelligence collection and its swift dissemination to police officers in the field could have played a major role in frustrating conspiracies. Further, even where intelligence is available, many agencies fail to collate the available inputs into an understandable whole for an effective follow-up by field operatives. This was, in essence, the conclusion of the 9/11 Commission appointed by President Bush immediately after the tragedy.

10.1 The U.S. Experience

Headed by former New Jersey Governor Thomas Kean, the National Commission on Terrorist Attacks Upon the United States (i.e., the 9/11 Commission) examined nearly 1,100 witnesses before submitting its comprehensive report, *the Final Report of the National Commission on Terrorist Attacks Upon the United States* (2004). Among its conclusions, the 9/11 Commission called for integrating the work of the various intelligence agencies to make counter-terrorism more effective. One principal recommendation was the creation of a National

[1] This situation, of mistakenly releasing terrorist leaders from custody is not unique to India. Al-Qaeda leader Ayman al-Zawahiri was held in a Russian jail from December 1996 to May 1997 without the Russian authorities realizing who was in their custody (Wright 2002).

Counter-terrorism Centre (NCTC) that would draw its staff from multiple departments and coordinate between agencies in the twin tasks of analyzing and disseminating intelligence pertaining to terrorism, as well as to plan operations that would strike before a terrorist group could carry out its plans. The 9/11 Commission Report boldly stated its vision for the NCTC. Breaking the older mold of national government organizations, this NCTC "should be a center for joint operational planning *and* joint intelligence…" (The 9/11 Commission 2004).

The NCTC came into existence through the Intelligence Reform and Terrorism Prevention Act 2004 (IRTPA). Its mission statement reflects the vision of its founders: "(To) (l)ead our nation's effort to combat terrorism at home and abroad by analyzing the threat, sharing information with our partners, and integrating all instruments of national power to ensure unity of effort" (NCTC.gov 2013).

The diversity of its staff, both analysts and planners, are drawn from the various components of the U.S. intelligence community and has been its main strength. The creation of a Pursuits Group in 2010 focusing on information that might uncover threats to U.S interests abroad has been a significant step forward in enhancing NCTC's reputation (Olson 2011). The fact that the U.S. has not suffered a *major* terror attack since 9/11 is at least partially a testament to NCTC's effectiveness.

10.2 The Indian Experience

The assault on Mumbai of November 26, 2008 (generally referred to as 26/11) took 166 lives and shocked the entire world and especially the population of Mumbai. This was the third time Mumbai was struck by a mega-terror attack.[2]

26/11 triggered a major public debate in India over the inadequacies of the Mumbai City Police and on the gaps in the Indian intelligence apparatus. There was a blame game immediately after the tragic events between the principal national intelligence organization, the Intelligence Bureau (IB), attached to the Ministry of Home Affairs (MHA) of the Central government in New Delhi, and the State Police. The State Police have claimed that the inputs received from the IB were not helpful.

The IB is a low-key organization with a rich tradition of professionalism (Datta 2006). It has never sought publicity or engaged in the aggressive over-zealous techniques of the sort that its counterparts display elsewhere in the world.[3] Recently, in a marked change in style from their conservative predecessors, some IB chiefs have become more technology-savvy and assertive in carrying out the Bureau's mission. This is a shift in organizational culture. The rise of a constant

---

[2] In 1993 a series of explosions masterminded by the Dawood Ibrahim killed 257 people, and in 2006 a series of bombs planted on Mumbai 's commuter trains by IM and LeT killed 206 people.

[3] One of the authors of this book had a stint in the IB for more than a decade.

terrorist threat has spurred this dramatic change in the IB. Unfortunately, the IB's failures, more than its triumphs, have grabbed public and media attention. Nevertheless, successive governments have relied heavily on the IB for its assessments of the state of internal security across the country. No one in the government has expressed open dissatisfaction with its performance.

The task of protecting the integrity of India and the lives of its citizens is complicated by two factors:

- the presence of a hostile neighbor, Pakistan on the northern and western frontiers; and
- the provision in India's Constitution that gives nearly absolute authority to the 28 States in matters of policing.

While the former enlarges the dimensions of governmental responsibility, the latter complicates it because the States are extremely sensitive to any attempt by the Federal government to advise them on issues involving maintenance of law and order. Coordinating measures to preserve internal security is a major challenge for the Federal government. In response to this challenge in 2002, the government created a Multi-Agency Centre (MAC) in New Delhi along with several regional centers (Gordon 2010). The rationale for this move was provided by the 2001 Report of a Group of Ministers tasked to undertake a thorough examination of the internal security apparatus. After an exhaustive probe, the report concluded that in view of internal security threats arising from extremists and separatist movements with support from outside India, increasing co-ordination between the Central and State Intelligence Agencies (or State Special Branches) as well as para-military intelligence branches was imperative. The report finds that these organizations "must not only formulate intelligence priorities, needs and requirements, but also devote special attention to streamlining intelligence operations." The report proposes a permanent Joint Task Force on Intelligence (JTFI), where the IB would take the lead role, along with the MHA and the representatives from the intelligence branches of the concerned States and Central Para Military Forces (CPMFs). The JTIF would determine intelligence priorities, intelligence needs, intelligence requirements and training facilities across the country and would assign specific responsibilities to the appropriate organizations. Such permanent centralization would move towards the collection, processing and analysis of intelligence data across multiple levels and agencies (Natarajan 2013).

The MAC has been a crucial element in forging a partnership between the Federal and State levels, finding national acceptance despite some early cynicism among some State governments. Besides the daily exchange of intelligence, State intelligence bureaus have found frequent participation in MAC meetings valuable, thus establishing a positive relationship of mutual trust between the IB in Delhi and the Special Branch CID among the States. An objective research study, beyond the scope of this work, would be needed to establish whether this arrangement has gone beyond cooperation and has actually helped to prevent terrorist attacks (Gordon 2010; India Today 2009).

The then-new Home Minister, Harvard-educated P. Chidambaram, who was appointed after 26/11, called for even greater reforms in India's approach to counter-terrorism, including the establishment of a National Counter-Terror Center. Chidambaram readily learned from the experiences of other countries, especially the United States. His speech, "A New Architecture for India's Security," given on the occasion of IB's 22nd Centenary Endowment Lecture, is a landmark, because, for the first time, a senior government functionary outlined the need for an NCTC. He did not refer to the Multi-Agency Centre (MAC) that was already in place, nor did he explain how an NCTC was an improvement over the MAC. However the Home Minister made it clear that the NCTC was an imperative in the context of the Mumbai attack and was therefore his top priority. He went on to indicate that the end of December 2010 was his deadline for the creation of this new organization (Chidambaram 2009). In a harbinger of the political and organizational challenges the creation of the NCTC would face, the Cabinet Committee on Security did not clear the proposal until January 2012 (Fair 2012).

10.3 NCTC-India Since 2012

Although there was no major opposition to the creation of a MAC and the latter, in general, has been functioning smoothly, the response of States to the proposed NCTC has been hostile. (The MAC could be an integral part of the proposed NCTC). The chief ministers of at least seven states have categorically opposed establishing the NCTC, and many others have expressed strong reservations. This hostility was because the NCTC, as conceived by the MHA would have conferred upon IB the authority to make *suo motu* arrests (i.e. arrests without approval by another authority such as the State Police). Repeated pronouncements by many State Chief Ministers point to a distrust and suspicion of New Delhi's intentions (Raman 2012; Rediff.com 2012).

States are jealous of inroads into their own authority and would not approve of any attempt by others to reduce their sphere of dominance. The IB has never enjoyed any legal authority in the past, being a mere "Attached Office" of the MHA without statutory recognition. Now, for the first time, a proposal has been mooted to dramatically increase the IB's power. States saw this as an attempt to erode their authority and were unwilling to approve the NCTC in the form it was presented to them by the Federal government. Several meetings and rounds of discussions by the MHA with State Chief Ministers have not borne fruit. Sensitive to the States' prerogatives, the MHA has now formulated a new scheme for the NCTC, conceding two points demanded by the States:

- The NCTC will be outside the IB and function directly under the MHA.
- The NCTC will not exercise any authority (e.g., making arrests) without the knowledge of the State concerned. The Police Chief in each State, will be the

nodal person for being kept informed by NCTC whenever any major action against terrorism is contemplated (Mohan 2013).

One possibility that might help address these issues of concern to the States is greater integration with the National Investigation Agency (NIA), which is part of the Ministry of Home Affairs. The NIA already coordinates widely across the various security agencies within India and also coordinates with law enforcement organizations in many different States. As a consequence, an NCTC might function as a part of MHA via a structure that leverages the strengths of the NIA.

Another possible organizational structure for an NCTC could be as an initiative of India's National Security Council with joint governance via a management board consisting of representatives from state police organizations as well as the Intelligence Bureau, Research & Analysis Wing, and other security organizations represented in the National Security Council.

10.4 The Future of NCTC-India

As this book is going to press, the establishment of the NCTC remains incomplete. Many of the States oppose even the revised formula for the NCTC advocated by the MHA. The trust deficit between the two sides is too large for a quick resolution. There are also huge political differences between the ruling coalition at the Centre and the various parties that control the States (PTI 2013). Lastly, the national level intelligence agencies face bureaucratic divides that also hamper the counter-terror mission (Swami 2013).

But the political difficulties in establishing an Indian NCTC do not reduce the continuing need for one. *The findings of our model of the behavior of IM highlight the need for more effective gathering, analysis, and dissemination of intelligence to better predict and prevent IM terror attacks*. To reiterate just a few of the examples, to better predict attacks, Indian security forces need better information about when IM operatives meet and link with other terrorist groups. To gather this kind of data requires the integration of travel, financial, signals, and human intelligence. This can best be achieved in an inter-agency bureau that incorporates experts from a range of agencies. As another example, the arrest of IM operatives is a predictor that more attacks are in the works. To increase the likelihood of preventing these attacks, Indian security needs the ability to rapidly combine whatever intelligence is gained from these arrests with the existing database and disseminate these findings in real time throughout India's national, state, and local security apparatus. *An NCTC would be well placed to take on these tasks*.

Since the rise of international terror networks, it has become a maxim that "it takes a network to beat a network." India faces sophisticated networked opponents. NCTC would be an essential node in a much-needed counter-network.

References

Chidambaram, P (2009) A new architecture for India's security. Intelligence Bureau Centenary Endowment Lecture, December 23. http://static.indianexpress.com/frontend/iep/docs/Chidambaram-speech.pdf

Datta S (2012) DNA special: intelligence bureau to take off cloak and bare the dagger. *DNA India* July 30. http://www.dnaindia.com/india/1721520/report-dna-special-intelligence-bureau-to-take-off-cloak-and-bare-the-dagger

Fair CC (2012) Prospects for effective internal security reforms in India Commonwealth and comparative politics 50:145-170

Gordon S (2010) India's unfinished security revolution. Institute for Defence Studies and Analysis Occasional Paper 11 August

Indian Express (2013) Yasin Bhatkal walked out of Kolkata jail as Bulla Mallik. Indian Express February 24. http://www.indianexpress.com/news/yasin-bhatkal-walked-out-of-kolkata-jail-as-bulla-mallik/1078815/.

India Today (2009) Are we any safer? India Today November 19. http://indiatoday.intoday.in/story/Are+we+any+safer/1/71557.html

Mohan V (2013) Govt not keen on making more changes to NCTC. Times of India June 5. http://timesofindia.indiatimes.com/india/Govt-not-keen-on-making-more-changes-in-NCTC/articleshow/20436478.cms

Natarajan S (2013) Recommendations on internal security,. Lok Sabha Question. March 19. http://mha.nic.in/par2013/par2013-pdfs/ls-190313/3615.pdf

Olson M (2011) Hearing before the Permanent Select Committee on Intelligence. U.S. House of Representatives October 6. http://www.nctc.gov/press_room/speeches/dnctc_testimony_before_hpsci_111006.pdf

PTI (2013) Centre, states need to work together for NCTC: home minister Sushilkumar Shinde, Press Trust of India June 11. http://economictimes.indiatimes.com/news/politics-and-nation/centre-states-need-to-work-together-for-nctc-home-minister-sushilkumar-shinde/articleshow/20542335.cms

Raman B (2012) The NCTC controversy. Outlook India March 5. http://www.outlookindia.com/article.aspx?280150

Rediff.com (2012) No consensus at NCTC meet; need anti-terror body, says PC. Rediff.com May 5. http://www.rediff.com/news/report/no-consensus-at-nctc-meet-need-anti-terror-body-says-pc/20120505.htm

Swami P (2013) Five years after 26/11, India faces intelligence famine. The Hindu February 27. http://www.thehindu.com/news/national/five-years-after-2611-india-faces-intelligence-famine/article4456325.ece

The 9/11 Commission (2004) The final report of the national commission on terrorist attacks upon the United States. July 22. http://www.9-11commission.gov/

Wright L (2002) The man behind Bin Laden. The New Yorker September 16

References

Appendix A
Data Methodology

This appendix describes the methodology used in our study of IM. The four-step process is explained below. Similar forecasts made regarding LeT have already proven true.[1]

Step 1 Systematically Gather Data

Data were gathered by studying a wide variety of open source literature. A total of 770 variables were identified. In general, datasets of this nature focus on a particular aspect of a group's behavior and are shaped by the interests and methods of the particular discipline studying an issue (e.g., databases assembled by economists focus on economic factors) and test particular hypotheses. This study is not constrained by anyone discipline and includes a broad range of variables covering economic, social, cultural, and organizational factors in the group's behavior.

The dataset consists of two types of variables: environmental variables and action variables. Environmental variables include variables describing the organizational structure and behavior of IM along with aspects (e.g., social, political, religious, financial, and ethnic) of the environment in which IM is functioning during a given month. Environmental variables also capture, through specific variables, the behaviors of other actors (e.g., the Pakistani government, the Pakistani military, the USA, and India) whose acts may impact the behavior of IM. The second set of variables are action variables that describe various aspects of the violent actions taken by IM during a given month.

It includes a quantitative value for each variable (e.g., number of people killed, number of attacks on professional security forces, and number of bombings).

The IM dataset captures information on a monthly basis, starting from January 2003 to September 2010; analyses in this book is based on that data. The data however have been augmented in various parts of this book to also include data extending through April 2013.

[1] This appendix is based on Appendix A in the book by V. S. Subrahmanian et al. (2012) but with minor variations and updates.

V. S. Subrahmanian et al., *Indian Mujahideen*, Terrorism, Security, and Computation,
DOI: 10.1007/978-3-319-02818-7,

Step 2 Uncover Behavioral Models

The second step in the analysis is the use of data mining algorithms to automatically uncover models of IM's behavior. While extremely successful companies like Amazon and EBay use data mining on a regular basis, both governments and social science researchers have been slow to embrace the power of data mining, instead frequently arguing that reasoning about terrorist groups is too idiosyncratic a task for computational tools.

The IM dataset consists of 770 variables. Data have been collected on a monthly basis from January 2002 to September 2010. This creates a spreadsheet with 127 rows and 770 columns, and nearly 100,000 individual cells. For reference purposes, Microsoft Excel brings up approximately 15 columns on a single screen. For an analyst to even *visually* see the IM data would require a screen capable of accommodating 770 rows, which would require a computer screen about 50 times as wide as the typical screen. While a human analyst cannot hope to systematically analyze data of this magnitude, data mining tools can do so in real time and potential reveal links between variables that would not have been discovered through traditional analysis.

The data mining paradigm developed is the *Stochastic Opponent Modeling Agents (SOMA)*. The SOMA-rule Learning Algorithm automatically reveals SOMA models of terrorist groups and has already been applied to uncover behavioral models of several terrorist groups including Hamas and Hezbollah (Khuller et al. 2007; Mannes et al. 2008). The algorithm considers every "bad" act (e.g., targeting security infrastructure, targeting civilian infrastructure, and carrying out fedayeen attacks) and automatically uncovers logic conditions with regard to the environmental variables that neatly distinguish between when the group performs the act from when it does not. These logic conditions then define probabilistic rules that project when IM will carry out a particular act during a specific time frame or under some set of real or hypothetical circumstances. SOMA-rules are "if-then" rules that analysts and policy experts can easily understand and explain.

Section B of Appendix A describes what SOMA-rules look like and the kinds of statistical conditions SOMA-rules should satisfy in order to be considered "good." In this book, however, we have built upon SOMA-rules and instead use more sophisticated "temporal probabilistic" rules or TP-rules as described in Chap. 3.

Step 3 Create Possible Policies

Over the years, multiple systems have been developed to automatically generate policies. The TOSCA system (Parker et al. 2011) is very fast. An analyst can "feed in" a database (such as our IM dataset) and specify a goal he wants to achieve (e.g., simultaneously reduce attacks by IM on tourist sites to a certain number or less and reduce attacks on Indian security installations to a certain number or less). The system interprets this to mean that these two types of behaviors by IM must be reduced to the specified level without increasing the number of other IM acts. TOSCA tries to find ways in which the environmental variables can be "reset" so that the probability of achieving the user's stated goal is maximized and it allows the policy analyst to specify constraints.

Some of the environmental variables may serve as policy levers that the analyst can try to technically manipulate to increase the probability of desired results. Some levers may be impossible to use—and the analyst can specify these via constraints. Other levers may have high costs—and the analyst can specify these too via cost specifications.

Such a tool helps the analyst leverage the enormous power of modern computing technology to analyze very large multidimensional spaces (in this case, a 770-dimensional space) bringing his own knowledge (via his specification of constraints) to bear as an important input to the problem.

The Policy Analytics Generation Engine (PAGE) (Simari and Subrahmanian 2010) system performs a similar task as TOSCA, but uses the SOMA-rules generated directly. In this analysis, however, we chose a simpler method based on Bell (1994a, b) described in Chap. 10.

The remainder of Appendix A covers the first two of these four steps (and their associated software components) in greater detail. The forecasting methodology and the policy generation methods are described in Chaps. 3 and 10 respectively.

A.1 Systematically Gathering Data

This section provides a detailed overview of the IM dataset, which contains about 770 variables that can apply to any terrorist group. These variables have already been used to gather data about several terror groups including Jaish-e-Mohammed (JeM), the Student Islamic Movement of India (SIMI), and Indian Mujahideen (IM) in South Asia region, as well as the Forces démocratiques de libération du Rwanda (FDLR). Moreover, the *Automated Coding Engine* (ACE) has been recently developed to gather data on these 770 variables for any group by crawling Lexis-Nexis news feeds and automatically extracting the values of these variables from them. ACE is about 75 % accurate. While it needs human intervention to raise its accuracy, ACE yields significant savings in coding time. For this study of IM, however, the data were collected manually.

A.1.1 Sources

The data were collected by consulting a wide range of sources. Every effort was used to ensure the sources were authoritative and the information accurate. The sources include:

- International news sources
- National, regional, and local news sources
- Books, papers, and reports authored by various researchers

- Reports and transcripts from authoritative sources
- Selected web sites that track terrorism-related information.

All data for this project were gathered *manually* from these sources and are deemed correct by the authors.

A.1.2 Data Organization

The IM dataset is organized as a relational database table. The rows of this table correspond to months, starting in January 2003 and going up to September 2010. There are 93 months of data in this dataset. The columns of this table correspond to individual variables. As mentioned earlier, the variables fall into two categories. *Environmental* variables describe IM's structure, leadership, and communications along with social, cultural, ethnic, political, and financial aspects of IM's environment. They include actions taken by other actors which might have had an impact on IM's behavior. *Action* variables describe the types of terrorist actions taken by IM during these 93 months.

A.1.3 Environmental Variables

We studied 570 environmental variables. These variables fell into broad categories, listed below in alphabetical order.

Communication: This set of variables describes IM communications, including the nature of its message and the infrastructure IM used to propagate this message. Specifically, the variables describe the:

- *Addressee* of various communications: who was the intended audience, separated out by entity type (e.g., government vs. security forces vs. international organizations), by region (e.g., global audience vs. national audience), ethnicity (were communications directed at specific ethnicities or groups holding specific beliefs).
- *Medium* of the communications: Did IM use blogs, emails, web sites to get their message out? Did they use books, periodicals and conferences? Did they enlist clerics? Did they use social networking sites or radio or phone to get their message out?
- *Messaging* in the communications: Did IM call for violence, recruit for their cause, or call for a change in lifestyle of the population? Did IM claim of responsibility for attacks, publicly identify enemies, make confessions, or justify violence?

Structural: This set of variables describes the social, political, geographic, and economic milieu in which IM operates. Specifically, the variables describe:

- *Relationship with Security Forces*: these describe the behavior of security forces in areas where IM is active.
  - Bombardment
  - Burning settlements
  - Corruption
  - Imposing curfews
  - DDR (demobilization, disarmament and reintegration) programs
  - Desertion among security forces
  - Executions (of offenders)
  - Imposing restrictions on freedom of restriction
  - Shutting down public sites
  - Sealing off regions
  - Paying soldiers on a regular basis
  - Engaged in repression against civilians
  - Engaged in sexual violence against civilians
  - Providing social services
  - Suppression of the opposition
  - Engaged in torture.
- *Government's International Relations*: These variables capture the types of international relations the government was subject to or was engaged in.
  - Border closures/disputes
  - Allegations of war crimes against the government
  - Breaking or re-establishing diplomatic relations
  - International embargoes against the government
  - Whether the government receives foreign aid
  - Whether any government assets were frozen
  - Whether the government has been accused of manipulating humanitarian aid
  - Whether the government is engaged in international disputes or territorial disputes
  - Whether there are travel bans against government officials.
- *Government Legitimacy*: These variables relate to the legitimacy of the governments involved, including whether the governments were autocratic, had coup d'états, or came to power through legitimate elections.
- *Ecological aspects*: These variables relate to the ecological environment in which IM functioned such as the state of deforestation, availability of drinking water, and occurrence of natural disasters.
- *Physical Infrastructure*: These variables describe the physical infrastructure of the countries (basically India) where IM operates, such as the presence of airports, oil and gas facilities and pipelines, border controls, ports and harbors, railways, roadways, and waterways.
- *Physical Resources*: These variables describe the availability of natural and man-made resources in the region. The natural resource variables focused on availability of various minerals and ores, while the man-made resources focused

largely on arms and ammunition, automobiles and transportation vehicles, electricity/power, and medicines/pharmaceuticals.

- *Terrain*: These variables describe the kinds of terrain in which IM was operating.
- *Demographics*: These variables relate to the demographics of the population, such as fertility rates, life expectancy, male-female ratios, median age, and total population.
- *Economic Variables*: These variables relate to the resources of revenue for the countries involved, e.g., agriculture, fisheries, and industry. These variables are connected to food prices, GDP and per capita income, balance of revenues and expenditures, the inflation rate, and the poverty rate.
- *External Actor*: These variables describe the behaviors of external actors in the region including the presence of:
  - Foreign militaries
  - Foreign non-state armed groups
  - Government organizations
  - International businesses
  - Foreign international organizations
  - Peacekeeping forces
- *Security Related*: These variables cover the security situation in the region responding to issues such as:
  - Inter-group tension
  - Conflicts with casualties on the side of foreign security forces
  - Conflicts with a national security force with casualties on the security force side
  - Inter-clan/tribe tensions
  - NSAGs
- *Environment Social Structure*: These variables capture the social structure of the region's population, including distribution of clans/tribes, distribution of ethnicities and forms of belief, the existence of internally displaced persons, whether the population has access to the Internet, the presence or absence of media organizations, literacy rates, and the refugee populations.
- *IM Environmental Variables, Group Structure and Organization:* We defined a large set of variables to study IM itself.
  - *Basic Group Variables:* dealt with whether the group had multiple names, the size of the organization, whether the group dissolved or split during a time frame, and information on umbrella organizations with which the group is affiliated.
  - *Group Character Variables:* dealt with whether the group is militant, political, or religious.
  - *Group Equipment Variables:* deal with the types of armaments, vehicles, chemicals, and other equipment the group has in its possession.
  - *Group Infrastructure Variables:* deal with how the group functions. This includes whether the group is engaged in

Businesses
Charities
Commodities
Property businesses.

In addition, this class of variables includes information on whether the group has a network structure, and whether it maintains offices and training camps.

- *Intra-Organizational Conflict*: This set of variables examine whether the group was engaged in internal conflict and the cause of such conflict (for ideological reasons, due to differences in goals of factions, due to differences about leadership, or conflicts about resources).
- *Group Leadership*: This class of variables relates to the type of leadership of IM and whether those leaders were arrested, or released from arrest, whether they are charismatic, or whether they are corrupt. There are also variables related to whether the leadership is paid, whether they have military experience, whether they are spiritual, and so forth.
- *Legitimacy of Group Leadership*: These variables examine the sources of the IM leadership's legitimacy.
- *Group's Local Organization*: This set of variables deals with how the group operates geographically, such as whether or not they are a cross-border organization, operate in displaced persons' camps or not (both internally and transnationally), and whether or not the organization's areas of operation have expanded or contracted.
- *Group Membership*: This class of variables studies the membership of the group, and the types of populations from which the group recruits its members. It defines variables related to members' belief system, whether child soldiers and forceful recruitments are used, whether foreigners are join the group, whether IM uses gender as a basis to decide whom to recruit, whether members of IM are also members of other non-state armed groups, and the socio-economic status of recruits.
- *Group Organization Split*: This set of variables examines whether the group has split, why, and whether it has reunited.
- *Group Support*: This set of variables checks whether IM provided financial, material, military, or political support to another NSAG.
- *Group Aspirations and Objectives*: This set of variables seeks to describe the goals and aspirations of the group.
- *Group Alliances*: This set of variables looks at the types of relationships the group has with other actors:

Foreign governments
International businesses
NGOs
Other NSAGs
Security forces
Prominent leaders

- *Group Domestic Relations*: These variables examine the group's domestic relations with the local (or national) government— if IM members receive amnesty, if its members subject to arrest (or released from arrest), if arrest warrants, government bans, frozen assets were associated with the group, if the government mistreat group members or kill their members arbitrarily, if the government carry out raids and other forms of repression, and finally, the kinds of media statements put out by the government.
- *Group International Relations*: These variables describe IM's international relationships. This includes international allegations of human rights abuses/war crimes by IM, arrests/extraditions of IM members by foreign states, international bans, international designation of IM as a terror organization, international embargos, international arrest warrants, assets freezes by international parties, international sanctions and resolutions against IM, and travel bans on IM personnel. A further set of variables describes ongoing negotiations involving IM.
- *Sources of Support:* This set of variables describe support to IM from different entities. Support variables are defined both by the type of support (financial, military, political, and material) as well as by the entity providing the support (e.g., the local population, diaspora, local government, foreign state, intergovernmental organization, non-governmental organization, non-state armed group, etc.).

A.2 Extracting Behavioral Models

This section briefly describes what a *behavioral model* of IM looks like, and then explains how these behavioral models are automatically extracted from our IM dataset.

A *behavioral model* of a terror group consists of a set of probabilistic rules of the form, "When condition C is true (in the environment in which IM is operating), then there is a probability of P % that IM will carry out a given action A with intensity level I".

For example, we discovered the following SOMA-rule about IM[2]:

> When there is a warming of diplomatic ties between India and Pakistan, there is a 100 % probability that it will carry out bombings 5 months later. (Support = 5, Inverse Probability = 1, Negative Probability = 0.)

When we examine this SOMA-rule, we note that we can write it as the probabilistic logic rule given below.

$$bombings(1,1) : 1 \leftarrow warm\_diplomatic\_ties(1,1).$$

[2] Provided here solely as an example. Detailed discussion is provided in previous chapters.

All SOMA-rules have a *head* and a *body*. We illustrate these concepts via the example given below:

bombings (1, 1):1 is the *head* of this rule. The head of a rule always consists of:

- an action variable ("*bombings*" in this case),
- a range of values for the *bombing* variable—in this case, the range of values is [1, 1] indicating that this variable lies between 1 and 1 (i.e., equals 1), and
- A probability, which in this case is 1.

$warm\_diplomatic\_ties(1,1)$ is the *body* of this rule. In this example, the body consists of one "environmental atom" $warm\_diplomatic\_ties(1,1)$. In general, only environmental variables can be referenced in the body of a rule.

This rule says that the probability of the *bombings* variable having a value of 1, given that the $warm\_diplomatic\_ties(1,1)$ variable has a value in the 1–1 interval, is 100 %.

Thus, all SOMA-rules have the form

$$HeadVar(Val1, Val2) : Prob \leftarrow BodyAtom_1 \& \ldots \& BodyAtom_k.$$

In short, a general SOMA-rule may be read as "When $BodyAtom_1$ and $BodyAtom_2$ and $BodyAtom_k$ are all true in a given month, then there is a probability of *Prob* that *HeadVar* will have a value between *Val*1 and *Val*2 during that month."

SOMA-rules are extracted automatically using algorithms we have developed (Ernst and Subrahmanian 2009; Khuller et al. 2007). The question is, given a dataset such as our IM dataset, how do we decide what SOMA-rules are worth extracting? For this study of IM, our SOMA-rule extraction engine uses four parameters for any possible rule of the form

$$HeadVar(Val1, Val2) : Prob \leftarrow BodyAtom_1 \& \ldots \& BodyAtom_k.$$

The *Support* of this rule refers to the number of months when both the head and the body of this rule were true, i.e., when the atom in the rule head and all environmental atoms in the rule body were true. The SOMA-extraction engine requires that all rules have at least a minimal, user-specified support.

The *Confidence* (or *Probability*) of a rule refers to the ratio of the number of months when both the body and the head of the rule were true to the number of times just the body of the rule was true, i.e.,

$$Confidence = \frac{number\ of\ months\ when\ HeadVar(Val1, Val2) \& BodyAtom_1 \& BodyAtom_2 \& \ldots \& BodyAtom_k\ is\ true}{number\ of\ months\ when\ BodyAtom_1 \& BodyAtom_2 \& \ldots \& BodyAtom_k\ is\ true}$$

Thus, *confidence* of a rule is merely the conditional probability of IM taking an action at a certain intensity level, given that the body of the rule is true. The SOMA-rule extraction engine requires that all extracted rule pass a desired confidence threshold.

The *Inverse Probability* of a rule refers to the ratio of the number of months the head and the body of the rule are both true to the number of months the head is true. Thus, this inverse probability is the conditional probability of the body being true, given that the head is true.

$$InverseProb = \frac{number\ of\ months\ when\ HeadVar(Val1, Val2)\ \&BodyAtom_1\ \&BodyAtom_2\ \&\ldots\&BodyAtom_k\ istrue}{number\ of\ months\ when\ HeadVar(Val1, Val2)\ istrue}$$

The SOMA-rule extraction engine requires that the inverse probability exceeds a threshold. Intuitively, when both the *Probability* and the *Inverse Probability* are high, this means that the rule head and the rule body co-occur with a high degree of likelihood, thus implying a strong correlation between the two.

Finally, the *Negative Probability* of a SOMA-rule is the ratio of the number of months the head and the body of a SOMA-rule are both true to the number of months the body is not true. In other words, the negative probability of a SOMA-rule is merely the conditional probability of the head being true, given that the rule body is false.

$$NegativeProb = \frac{number\ of\ months\ when\ HeadVar(Val1, Val2)\& \sim (BodyAtom_1\ \&BodyAtom_2\ \&\ldots\&BodyAtom_k)\ istrue}{number\ of\ months\ when \sim (BodyAtom_1\ \&BodyAtom_2\ \&\ldots\&BodyAtom_k)\ istrue}$$

The SOMA-rule extraction engine only extracts rules that have a negative probability below a given threshold. This is because finding conditions that have a high probability of predicting an action (when they are true) and a low probability of predicting an action (when they are false) is the ultimate goal of these forecasting engines.

Finally, an important note on missing information: like any dataset, there is much missing information in our dataset. In order to address the issue of missing values in our calculations of Confidence, Negative Probability, Inverse Probability, and Support, we only considered months where data was present for *all* of the variables encoded in the conditions we were considering. For instance, if we are considering the confidence for the rule that 5 months after there is a warming of diplomatic ties between India and Pakistan, IM carries out a bombing, we would first count cases where there was a warming of diplomatic ties between India and Pakistan in some month m (i.e. the data was not missing) and there was an IM bombing in month $(m + 5)$. This would then be divided by the number of months in which there was a warming of diplomatic ties between India and Pakistan in some month m (i.e. the data was not missing) and there was data for the variable about IM bombing in month $(m + 5)$. In other words, missing data were not filled in by some assumption.

Appendix B
List of All Terrorist Attacks Carried Out by IM

Date	Location	Killed/injured	Description
1/22/02	Kolkata	4/20	Four gunmen on two motorcycles carried out drive-by shooting at American Center, killing four security personnel Attack claimed by ARCF and HuJI in revenge for Asif Reza's death; ARCF individuals later linked with IM (Battarcharya 2002 http://hindu.com/2002/01/23/stories/2002012305090100.htm)
8/25/03	Mumbai	52/200+	Two bombs made with military explosive RDX planted in two taxis detonated at the Gateway of India, where 16 killed; Zaveri Bazaar where 36 were killed No claim of responsibility, but LeT in combination with SIMI suspected Perpetrators motivated by atrocities against Muslims in Gujarat riots (Katakam 2003 http://www.frontline.in/static/html/fl2019/stories/20030926003802100.htm; Zee News 2009 http://zeenews.india.com/news/nation/all-three-let-terrorists-convicted-for-2003-mumbai-blasts_550599.html)
2/23/05	Varanasi	9/20+	Bombs planted at Dasahwamedh Ghat Initially police determined it was a cooking cylinder explosion; after 2006 Varanasi attack, police declared this a terrorist incident No claim of responsibility, SIMI and IM suspected IED pressure cooker bomb packed with either RDX or ammonium nitrate; a second IED did not detonate (Gupta 2011)
7/28/05	Jaunpur, UP	13/52	Bomb detonated in the unreserved compartment of the Shramjeevi Express headed to Delhi Initially explosion blamed on leaking gas cylinder, on HuJI IM operative Sadique Sheikh confessed that it was an IM operation Bomb was believed to be a pressure cooker containing RDX (Gupta 2011)

(continued)

V. S. Subrahmanian et al., *Indian Mujahideen*, Terrorism, Security, and Computation,
DOI: 10.1007/978-3-319-02818-7,

(continued)

Date	Location	Killed/injured	Description
10/29/05	Delhi	67/225	Three blasts in Delhi markets as people celebrated Diwali and Eid-ul-Fitr • 5:38 p.m., IED planted on a scooter detonated at Paharganj market on Chheh Tutu Chowk in central Delhi • 6 p.m., bomb in black bag detonated at Govindpuri market • 6:05 p.m., a car bomb detonated at Sarojini Nagar market, killing 43 as the explosion ignited cooking gas cylinders Islamic Revolutionary Front, previously unknown, claimed responsibility but police suspect LeT In 2008, IM operatives confessed to Delhi bombings (Gupta 2011)
3/7/06	Varanasi	16–28/ 101+	Two roughly simultaneous blasts • 6:20 p.m., Sankat Mochan temple, killing at least seven, injuring over 40 • Moments later, a second bomb detonated, in the waiting room of the railway station, killing at least nine and injuring about 50 • A third bomb did not detonate, discovered near the Gowdhulia market All three bombs were pressure cookers packed with RDX and ammonium nitrate, concealed in black bags Claimed by Lashkar-e-Qahar, a little known Kashmiri group, but Indian police suspected a group of Bangladeshis In 2008, under interrogation, IM leader claimed that all three bombings was IM operation (Gupta 2011)
7/11/06*	Mumbai	209/714	Seven bombs detonated on Mumbai commuter trains at 6:24 p.m. Bombs were pressure cookers packed with RDX, stowed in the overhead luggage racks of the trains Initially police suspected LeT; in 2008 and 2010 arrested IM figures stated it was an IM operation Executed with support from LeT and possibly ISI Blasts de-railed Pakistani-Indian counter-terror cooperation (Gupta 2011; (Tankel 2011)
2/18/07	Panipat	68/50	2 bombs exploded on Samjhauta Express (twice-weekly train between Delhi and Lahore) 68 people, mostly Pakistani nationals, were killed IM suspected, as are LeT and Hindu extremist group Abhinav Bharat (Gupta 2011)
5/22/07	Gorakhpur	0/0	Three ammonium nitrate bombs concealed in milk jugs, planted in market, at market exit, and at electrical transformer Failed to detonate properly, no injuries (Gupta 2011)

(continued)

(continued)

Date	Location	Killed/injured	Description
8/25/07	Hyderabad	42/54	Two near simultaneous bombs detonated, first at a popular eatery, 32 killed, and then in an auditorium at an amusement park, 10 killed A third bomb, targeting a park, did not detonate Bombs contained explosives made from ammonium nitrate and fuel oil IM operative revealed IM's responsibility under interrogation years later (Gupta 2011; Jafri 2007 http://www.rediff.com/news/2007/aug/25hydblast.htm)
11/23/07	Varanasi, Faizabad, Lucknow	14/57–81	Bombs planted on bicycles outside of courthouses in three cities • 1:05 p.m., Varanasi, two bombs detonated, 9 killed • 1:05 p.m., Lucknow, one bomb detonated, another failed to do so, none killed • 1:15 p.m., Faizabad, two bombs detonated, five killed Bombs were pressure cookers packed with ANFO and ball bearings Police initially suspected HuJI 1:00 p.m., email sent to TV news channels claimed IM responsibility • stated lawyers were targeted because in Lucknow lawyers had beaten accused JeM operatives who had been targeting Rahul Gandhi and Uttar Pradesh bar association had refused to defense Mohammed Walliullah who was accused in 2006 Varanasi blast • stated bombings were revenge for 1992–1993 and 2002 attacks on Muslims • denied any connections with ISI, LeT, HuJI or other terrorist groups (Gupta 2011; Sahu 2011 http://www.indianexpress.com/news/huji-or-im–case-on-but-two-theories-in-up-court-blasts/843326/0)
5/13/08	Jaipur	80/216	7:00–7:35 p.m., nine bombs detonated in seven locations around marketplaces in Jaipur Bombs were made with ANFO IM claimed responsibility over an email, police blamed HuJI (Gupta 2011)
7/25/08	Bengaluru	2/7–8	1:20–2:35 p.m., seven low intensity blasts on crowded roadsides, bus stops, and shopping centers throughout the city; eighth bomb was discovered and defused Ammonium nitrate bombs did not detonate properly, limiting casualties LeT claimed responsibility and may have funded the attacks (Gupta 2011; Times of India 2008 http://articles.timesofindia.indiatimes.com/2008-07-25/india/27892409_1_bomb-blasts-bangalore-blasts-fifth-blast)

(continued)

(continued)

Date	Location	Killed/injured	Description
7/26/08	Ahmadabad	57/100+	6:15–7:00 p.m., about 20 low-intensity bombs detonated all over the city 7:30 p.m., a car bomb made from ANFO and an LPG cylinder detonated next to City Trauma Centre, 37 killed Security forces defused two car bombs and 23 smaller bombs which had malfunctioned 6:10 p.m., IM emailed claim of responsibility (Gupta 2011)
9/13/08	New Delhi	30/130	Five blasts in crowded areas • 5:55 p.m., first blast, Gaffar Market, 20 killed • Over next 35 min, four low-intensity blasts, Connaught Place and M-Block Market • Bomb near Children's Park discovered and defused 6:27 p.m., IM email claim sent from hacked account Bombs made from ammonium nitrate and ball bearings Sept. 19, Batla House Encounter, police raided IM safe house, captured one, killed two and obtained key documents (Gupta 2011)
10/30/08	Assam	83/300–400	11 a.m. and subsequent, between 10–18 blasts in four cities Three car-bombs in Guwahati, 31 killed Attack claimed by ISF-IM via SMS to local television station HuJI and ULFA most likely suspects (Kalita 2008 http://articles.timesofindia.indiatimes.com/2008-11-01/india/27895609_1_guwahati-blasts-khagen-sarma-ulfa; Thottam 2008 http://www.time.com/time/world/article/0,8599,1855023,00.html)
2/13/10	Pune	17/54+	6:50 p.m., bombing at German Bakery frequented by foreigners Five deaths of the 17 deaths and 10 of the injured were injured foreign nationals Bomb was ammonium nitrate with RDX as a booster Joint LeT-IM operation LeT operative David Coleman Headley may have reconnoitered the target in 2008–2009 when he stayed at a hotel nearby (Gupta 2011; NDTV 2010c http://www.ndtv.com/article/india/did-david-headley-plan-pune-blast-16273)
4/17/10	Bengaluru	0/15	Two low-intensity blasts outside Chinnaswamy Stadium shortly before a cricket match; an additional bomb was found and defused (NDTV News 2010b http://www.ndtv.com/article/india/after-blasts-another-bomb-outside-bangalore-stadium-20219) Four IM activists later arrested for their involvement (PTI 2012 http://articles.timesofindia.indiatimes.com/2012-08-30/india/33498529_1_blast-case-chinnaswamy-stadium-bangalore-mumbai-ipl)

(continued)

(continued)

Date	Location	Killed/injured	Description
9/20/10	Delhi	0/2	Shooting by two motorcycle riding gunmen at Jama Masjid Mosque injures two Taiwanese tourists A car bomb failed to detonate IM emailed claim of responsibility, dedicating the attacks to the IM members killed a year earlier at Batla House and threatening to disrupt the Commonwealth Games which India was scheduled to host (Gupta 2011; Mid-Day 2010 http://www.mid-day.com/news/2010/sep/200910-Delhi-Indian-Mujahideen-terrorist-Jama-Masjid-Attack.htm)
12/7/10*	Varanasi	2/38	Bomb concealed in a milk tin exploded during sunset prayers at Shita Ghat killing two IM claimed the attack with an email minutes after explosion (Swami 2011 http://www.thehindu.com/news/national/article2436739.ece)
7/13/11*	Mumbai	27/127	Three bombs detonated • 6:54 p.m., bomb at Zaveri Bazaar • 6:55 p.m. bomb at Opera House district in a manhole • 7:05 p.m., bomb at a bus stop at Kabutarkhana IM operation with support from ISI, alleged that IM was a creation of ISI (BBC News 2011 http://www.bbc.co.uk/news/world-south-asia-14141454; Times of India 2012b http://articles.timesofindia.indiatimes.com/2012-06-02/india/31982947_1_bomb-blasts-chargesheet-anti-terrorism-squad),
9/7/11*	Delhi	13/80	Bomb detonated outside of Delhi High Court Emails from both IM and HuJI claiming responsibility, IM email threatened that another bomb had been planted at a shopping mall and claimed that HuJI had nothing to do with the attack (Times of India 2011a http://articles.timesofindia.indiatimes.com/2011-09-08/india/30129554_1_claims-responsibility-delhi-high-court-blast, 2011b http://articles.timesofindia.indiatimes.com/2011-09-07/india/30122504_1_huji-harkat-ul-jihad-delhi-high-court)
8/1/12*	Pune	0/1	Four low intensity blasts struck along the Junglee Maharaj Road at 7:30 p.m. Ammonium nitrate bombs, similar to 2008 bombings in Jaipur (Byatnal and Joshi 2012 http://www.thehindu.com/news/national/other-states/four-lowintensity-blasts-in-pune-one-injured/article3713166.ece; Sharma 2012 http://indiatoday.intoday.in/story/maharashtra-pune-blasts-indian-mujahideen/1/213378.html)

(continued)

(continued)

Date	Location	Killed/injured	Description
2/21/13*	Hyderabad	17/100+	Two bombs detonated in crowded Dilsukhnagar district 7:05 p.m., first bomb outside an eatery near a movie theater, the second at a bus about 5 min later Bombs were ammonium nitrate devices attached to bicycles National security agencies issued an alert due to the recent execution of Afzal Guru for involvement in 2001 Parliament attack Authorities suspected IM due to modus operandi and because IM operatives confessed to reconnoitering Dilsukhnagar in July 2012 (*Indian Express* 2013 http://www.indianexpress.com/fullcoverage/hyderabad-twin-blasts/487/)
4/17/13*	Bengaluru	0/16	Bomb strapped to a motorbike detonated 100 meters from BJP office at 10:30 a.m., most of the 16 injured were police Large crowd was present at the BJP office due to upcoming elections IM was an initial suspect (Zee News 2013 http://zeenews.india.com/news/nation/terror-strikes-bangalore-16-injured-in-bomb-blast-near-bjp-office_842911.html)

* Indicates that this attack was *not* included in the data used to model IM behavior. The reasons why an attack might not be included were that it occurred after the data collection was complete, or IM was not cited as a perpetrator during the period when data was being collected.

Appendix C
List of All Temporal Probabilistic Rules Presented in this Book

Rule Name	BOMB-1	BOMB-2	BOMB-3	BOMB-4	BOMB-5
Time Offset	2	2	2	4	2
Dependent Variable	IM carries out bombings	IM carries out bombings	IM carries out joint bombings	IM carries out bombings	IM carries out bombings
Lower Bound	1	1	1	1	0
Upper Bound	1	1	1	1	69
Support	5	5	4	4	4
Probability	100	100	100	100	100
Inverse Probability	100	100	100	100	100
Negative Probability	0	0	0	0	0
Independent Variable #1	IM personnel arrested	IM personnel arrested	IM communicating about its campaign	IM held a conference	IM made claims of responsibility
Independent Variable #2			IM shared members with other NSAGs		
Lower Bound	1	1	1	1	1
Upper Bound	21	21	1	1	1

V. S. Subrahmanian et al., *Indian Mujahideen*, Terrorism, Security, and Computation,
DOI: 10.1007/978-3-319-02818-7, © Springer International Publishing Switzerland 2013

Rule Name	BOMB-6	BOMB-7	BOMB-8	BOMB-9	BOMB-10
Time Offset	2	2	5	1	1
Dependent Variable	IM carries out bombings	IM carries out bombings	IM carries out bombings	IM carries out bombings	IM carries out bombings
Lower Bound	1	1	1	1	1
Upper Bound	1	1	1	1	1
Support	5	6	5	4	4
Probability	100	100	100	100	100
Inverse Probability	100	100	100	100	100
Negative Probability	0	0	0	0	0
Independent Variable #1	IM shared members with other NSAGs	IM shared members with other NSAGs	India and Pakistan entertain diplomatic links	Internal violence within India	Internal violence within India
Lower Bound	1	1	1	1	2
Upper Bound	1	1	1	1	15

Rule Name	PS-1	PS-2	PS-3	PS-4	PS-5
Time Offset	1	2	2	4	2
Dependent Variable	IM targets public sites	IM targets public sites	IM targets public sites	IM targets public sites	IM targets public sites
Lower Bound	1	1	1	1	1
Upper Bound	1	1	1	1	1
Support	4	4	4	4	5
Probability	100	100	100	100	100
Inverse Probability	100	100	100	100	100
Negative Probability	0	0	0	0	0
Independent Variable #1	IM personnel arrested	IM communicating about its campaign	IM made claims of responsibility	IM held a conference	IM shared members with other NSAGs
Lower Bound	1	1	1	1	1
Upper Bound	21	1	1	1	1

Rule Name	PS-6	SA-1	SA-2	SA-3	SA-4
Time Offset	5	4	4	2	2
Dependent Variable	IM targets public sites	IM carries out simultaneous attacks	IM carries out timed attacks	IM carries out timed attacks	IM carries out simultaneous attacks
Lower Bound	1	1	1	1	1
Upper Bound	1	1	1	1	1
Support	4	4	4	4	3
Probability	100	100	100	100	100
Inverse Probability	100	100	100	100	100
Negative Probability	0	0	0	0	0
Independent Variable #1	India and Pakistan entertain diplomatic links	IM held a conference	IM held a conference	IM personnel arrested	IM personnel arrested
Lower Bound	1	1	1	1	1
Upper Bound	1	1	1	21	21

Rule Name	SA-5	TK-1	TK-2	TK-3	TK-4
Time Offset	2	4	2	1	5
Dependent Variable	IM carries out timed attacks	Total killed in IM attacks	Total killed in IM attacks	Total killed in IM attacks	Total killed in IM attacks
Lower Bound	1	0	0	0	0
Upper Bound	1	69	69	69	69
Support	4	4	3	4	3
Probability	100	100	100	100	100
Inverse Probability	100	100	100	100	100
Negative Probability	0	0	0	0	0
Independent Variable #1	IM shared members with other NSAGs	IM held a conference	IM communicates about its campaign	IM personnel arrested by government	India and Pakistan entertain diplomatic links
Lower Bound	1	1	1	1	1
Upper Bound	1	1	1	21	1

Rule Name	TK-5	TK-6	TK-7	TK-8
Time Offset	2	2	1	1
Dependent Variable	Total killed in IM attacks	Total killed in IM attacks	Total killed in IM attacks	Total killed in IM attacks
Lower Bound	0	0	0	0
Upper Bound	69	69	69	69
Support	5	4	4	4
Probability	100	100	100	100
Inverse Probability	100	100	100	100
Negative Probability	0	0	0	0
Independent Variable #1	IM shared members with other NSAGs	IM made claims of responsibility	Internal violence within India with casualties	Internal violence within India
Lower Bound	1	1	1	2
Upper Bound	1	1	1	15

Appendix D
Instances of Improved Diplomatic Relations Between India and Pakistan

Date	Description	Source
May 2003	Pakistan and India re-establish formal diplomatic relations	http://www.globalsecurity.org/military/world/war/kashmir-2002.htm
August 2003	Indian officials refrain from blaming Pakistan for bombings in Mumbai; Pakistan condemns the bombings	http://www.nytimes.com/2003/08/26/world/50-dead-in-bombay-after-twin-blasts-in-crowded-areas.html
January 2004	India and Pakistan agree to composite dialogue process	http://www.idsa.in/idsastrategiccomments/Theriseoffiscalterror_SSaxena_250305
February 2004	India and Pakistan launch composite dialogue over all issues, including Kashmir	http://www.crisisgroup.org/en/regions/asia/south-asia/kashmir/B106-steps-towards-peace-putting-kashmiris-first.aspx
July 2005	Pakistani and India agree to continue talks after Ayodhya attack	http://news.bbc.co.uk/2/hi/south_asia/4654593.stm
October 2005	Indian PM and Pakistani President discussed recent bombings	http://news.bbc.co.uk/2/hi/south_asia/4390460.stm
December 2006	Former Pakistani President Musharraf states that India and Pakistan were close to an agreement	http://www.thenews.com.pk/latest-news/2638.htm
May 2008	Indian PM visiting Pakistan to review peace process	http://www.reuters.com/article/2008/05/13/idUSSP20219
May 2009	Responding to Pakistani requests, India provides more evidence regarding Mumbai attacks	http://news.rediff.com/report/2009/may/20/mumterror-india-hands-over-more-evidence-to-pakistan.htm
June 2009	Indian PM Singh and Pakistani President Zardari meet at a summit	http://news.bbc.co.uk/2/hi/south_asia/8102223.stm
July 2009	U.S. and India sign defense pact, Indian and Pakistani PMs meet	http://news.bbc.co.uk/2/hi/south_asia/8158489.stm

(continued)

V. S. Subrahmanian et al., *Indian Mujahideen*, Terrorism, Security, and Computation,
DOI: 10.1007/978-3-319-02818-7,

(continued)

Date	Description	Source
September 2009	Indian and Pakistani Foreign Ministers meet, follow-up to July meetings of Prime Ministers	http://news.rediff.com/report/2009/sep/21/india-pak-foreign-ministers-to-meet-sep-27.htm
November 2009*	Indian PM Singh travels to US, meets with U.S. president	http://news.rediff.com/report/2009/nov/25/safe-havens-of-terrorists-must-be-eliminated.htm
February 2010	Indian and Pakistani foreign secretaries meet to "restore trust"	http://www.indianexpress.com/news/indopak-talks-india-prioritises-terror-saeeds-arrest/584333/
March 2010*	Pacific Command chief describes growth in U.S.-Indian military ties	http://zeenews.india.com/news/south-asia/let-expanding-to-other-south-asian-nations_614460.html
April 2010	Indian and Pakistani PMs met at nuclear summit	http://news.bbc.co.uk/2/hi/south_asia/8616873.stm
May 2010*	India signed three bilateral cooperation pacts with China	http://news.rediff.com/report/2010/may/27/india-china-sign-three-bilateral-pacts.htm
June 2010	Senior diplomats of India and Pakistan met in Islamabad to improve relations and lay groundwork for meeting between foreign ministers on July 15, 2010	http://www.bbc.co.uk/news/10399327
July 2010	Foreign Ministers of India and Pakistan met in Islamabad for "constructive" and "useful" talks	http://www.thehindu.com/news/national/india-pakistan-decide-to-remain-engaged/article517384.ece

Appendix E
Instances of Indian Mujahideen Conferences

Date	Description	Source
July-August 2004*	Jolly Beach meetings establishing IM	http://www.ctc.usma.edu/posts/riyaz-bhatkal-and-the-origins-of-the-indian-mujahidin
January 2008	Meetings in jungles near Pavagadh where the Gujarat operation was planned	http://articles.timesofindia.indiatimes.com/2008-08-17/ahmedabad/27929682_1_simi-members-serial-blasts-safdar-nagori
March 2008	IM held several meetings at unspecified locations	http://timesofindia.indiatimes.com/Cities/Ahmedabad/Indian_Mujahideen_another_face_of_SIMI_/articleshow/3372572.cms
April 2008	End of April, Kalupur shop of Yunus Mansur (suspected IM members) where they decided to carry out July 26 attacks in Gujarat	http://timesofindia.indiatimes.com/Cities/Ahmedabad/Indian_Mujahideen_another_face_of_SIMI_/articleshow/3372572.cms
May 2008	At a lodge in Kuttypuram, Karnataka, Bhatkal met with Abdul Sattar, planned 2008 Bangalore bombings	http://news.rediff.com/special/2009/jul/24/confessions-of-a-bomb-maker.htm
June 2008	Further meetings between Bhatkal and Abdul Sattar in Kuttypuram to prepare for Bengaluru attack	http://newindianexpress.com/cities/bangalore/article99513.ece
August 2008	Bhatkal met with Abdul Sattar and others in Hyderabad, discussed operation and Bhatkal told Abdul Sattar to go underground	http://news.rediff.com/special/2009/jul/24/confessions-of-a-bomb-maker.htm
January 2010*	Yassin Bhatkal and Mohsin Choudhary met Himayat Baig to finalize Pune attack	Gupta *Indian Mujahideen* (2011)

* Conferences not included in the data collected and coded

V. S. Subrahmanian et al., *Indian Mujahideen*, Terrorism, Security, and Computation, DOI: 10.1007/978-3-319-02818-7, © Springer International Publishing Switzerland 2013

Appendix F
Indian Mujahideen Claims of Responsibility for Attacks

Date	Description	Source
July 2006	Media outlets received messages from IM claiming responsibility for bombings before November 2007, including Mumbai train bombings	http://ramansterrorismanalysis.blogspot.com/2009/09/continuing-threat-from-indian.html
November 2007	IM sent an emailed claim of responsibility to media outlets moments before bombs detonated at courthouses in Varansi, Lucknow, and Faizabad	http://www.reuters.com/article/2008/09/15/idUSDEL355542
May 2008	A day after Jaipur bombings, IM emailed claim of responsibility with video clip of bicycles used in the bombings	http://beta.dawn.com/news/303016/group-claims-carrying-out-jaipur-bombings
July 2008	Email claiming responsibility for Ahmedabad bombing was sent to media outlets 5 min before bombs detonated	http://www.atimes.com/atimes/South_Asia/JG29Df02.html
August 2008	Email claiming responsibility for July Ahmedabad bombing sent to media outlets that included pictures of cars used in bombing and claiming no one involved in the bombing had been arrested	http://www.satp.org/satporgtp/countries/india/terroristoutfits/simi_tl.htm
September 2008	Email claiming responsibility for five bombs across Delhi sent to media outlets 10 min before first blast	http://timesofindia.indiatimes.com/Cities/Delhi/Delhi_30_dead_as_5_bombs_go_off_in_45_min/rssarticleshow/3479914.cms
October 2008	Email claiming responsibility for Guwahati blasts of October 30, 2008 in the name of "Islamic Security Force (Indian Mujahideen)"	http://www.idsa.in/backgrounder/IndianMujahideen
February 2010	IM claimed responsibility for February 13 bombing of the German bakery in Pune	http://www.india-forums.com/news/national/231595-simi-indian-mujahideen-own-up-to-pune-blast-police.htm
September 2010	Email claimed responsibility for Jamia Masjid attack sent to media outlets $2^{1/2}$ h after attack	http://www.indianexpress.com/news/on-batla-anniversary-email-talks-of-martyrs-kashmir-genocide/684173/0
December 2010*	Email to newsrooms throughout India claiming responsibility minutes for bomb at Shita Ghat in Varanasi, moments after detonation	http://www.thehindu.com/news/national/article2436739.ece
September 2011*	Email claiming responsibility for bombings in Delhi sent the day after the blast and after HuJI claimed responsibility	http://articles.timesofindia.indiatimes.com/2011-09-08/india/30129554_1_claims-responsibility-delhi-high-court-blast

* Claims of responsibility were issued after coding project was complete and not included in the data coded

V. S. Subrahmanian et al., *Indian Mujahideen*, Terrorism, Security, and Computation,
DOI: 10.1007/978-3-319-02818-7,

Appendix G
Instances of Indian Mujahideen Communications About Its Terror Campaign

Date	Description	Source
May 2008	Email sent after Jaipur bombings explained attack was intended to send message to Indian Hindus and to undermine India's tourism industry	http://www.jihadwatch.org/archives/021063.php
July 2008	Email sent after Ahmedabad bombing hinted at further attacks and cited Muslim grievances	http://www.time.com/time/world/article/0,8599,1826950,00.html
September 2008	Email sent just before attack said IM was demonstrating its ability to strike in high security zones	http://www.outlookindia.com/article.aspx?238507
February 2010	IM threatens more serial bombings	http://zeenews.india.com/news/nation/nabbed-im-operative-was-planning-air-attacks_600855.html
September 2010	Email after Jamia Masjid attack warns of more IM attacks and states this attack was intended to embarrass India in the run-up to the Commonwealth Games	http://articles.timesofindia.indiatimes.com/2010-09-20/india/28237894_1_terror-attacks-fresh-attacks-al-arbi

References

Ernst J, Subrahmanian VS (2004) Method and system for optimal data diagnosis. U.S. Patent 7474987 January 6, 2009 Filed February 10, 2004

Gupta S (2011) The Indian Mujahideen: tracking the enemy within. Hachette, Gurgaon

Khuller S, Martinez V, Nau D, Simari G, Sliva A, Subrahmanian VS (2007) Computing most probable worlds of action probabilistic for logic programs: scalable estimation for 1,030,000 worlds. Annals of Mathematics and Artificial Intelligence 51:295–331

Mannes A, Michael M, Pate A, Sliva A, Subrahmanian VS, Wilkenfeld J (2008) Stochastic opponent modeling agents: a case study with Hezbollah. Proc 2008 First Intl Workshop Social Computing, Behavioral Modeling and Prediction. Springer Verlag, Phoenix

V. S. Subrahmanian et al., *Indian Mujahideen*, Terrorism, Security, and Computation, DOI: 10.1007/978-3-319-02818-7, © Springer International Publishing Switzerland 2013

Mannes A, Sliva A, Subrahmanian VS, Wilkenfeld J (2008) Stochastic opponent modeling agents: a case study with Hamas. Proc 2008 Intl Conf Computational Cultural Dynamics, AAAI Press

Parker A, Simari GI, Sliva A, Subrahmanian VS (2011) Approximate achievability in event databases. Proc ECSQARU 2011:737–74

Simari GI, Subrahmanian VS (2010) Abductive inference in probabilistic logic programs. Tech Communications of the 2010 Intl Conf on Logic Programming

Subrahmanian, VS, Mannes, A, Shakarian, J, Sliva, A, and Dickerson, J (2012) Computational analysis of terrorist groups: Lashkar–e–Taiba. Springer, New York

Tankel S (2011) Storming the world stage: the story of Lashkar-e-Taiba. C. Hurst & Co, London

Index

V. S. Subrahmanian et al., *Indian Mujahideen*, Terrorism, Security, and Computation,
DOI: 10.1007/978-3-319-02818-7, © Springer International Publishing Switzerland 2013

M

N

O

P

Q

R

S

T

U

Printed in the United States
By Bookmasters